Praise For The Dog With The Wind In Her Hair

A beautifully written book for all dog lovers and more..... definitely on my Christmas list!

D.H. - Amazon Kindle Customer ★★★★★

I thoroughly enjoyed this book. I feel like the author is a friend and we've enjoyed a satisfying conversation. I will read more of the author's books.

K.N. - Amazon Kindle Customer ★★★★★

With some sparkling prose, amusing and fascinating stories of her life with animals Pippa White's book is a must for dog lovers. However, there is more; the narrative shows the author's deep love of and affiliation with a range of creatures from ferrets to Guinea fowls and the importance of the natural world. In fact it is some of her descriptions of the passage of the seasons that have a lyrical quality that enchanted me.

Pam Keevil - Author of Virgin At 50 ★★★★★

Loved this book - so cleverly written and such a fantastic insight into life with animals and how they can literally change your life.

A-M.H. - Amazon Kindle Customer ★★★★★

Just what I've thought and felt but never bothered to write down.

H.P. - Amazon Kindle Customer ★★★★★

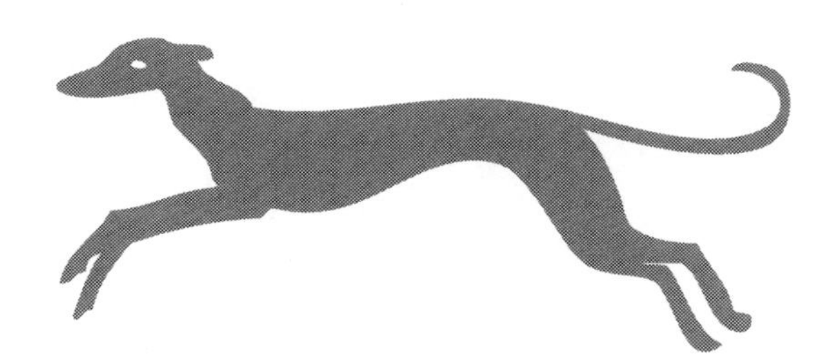

The Dog With The Wind In Her Hair

Pippa White

Ellie White
10 Hillesley
Wotton-Under-Edge
Gloucestershire
GL12 8RG
www.booksaboutdogs.co.uk

First published in Great Britain in 2019 by Ellie White.

Cover design and typesetting by Cotswold Websites.

www.cotswoldwebsites.co.uk

A CIP catalogue record for this book is available from the British Library.

First Published In Paperback In Great Britain In March 2019

ISBN: 978-1-909972-08-7

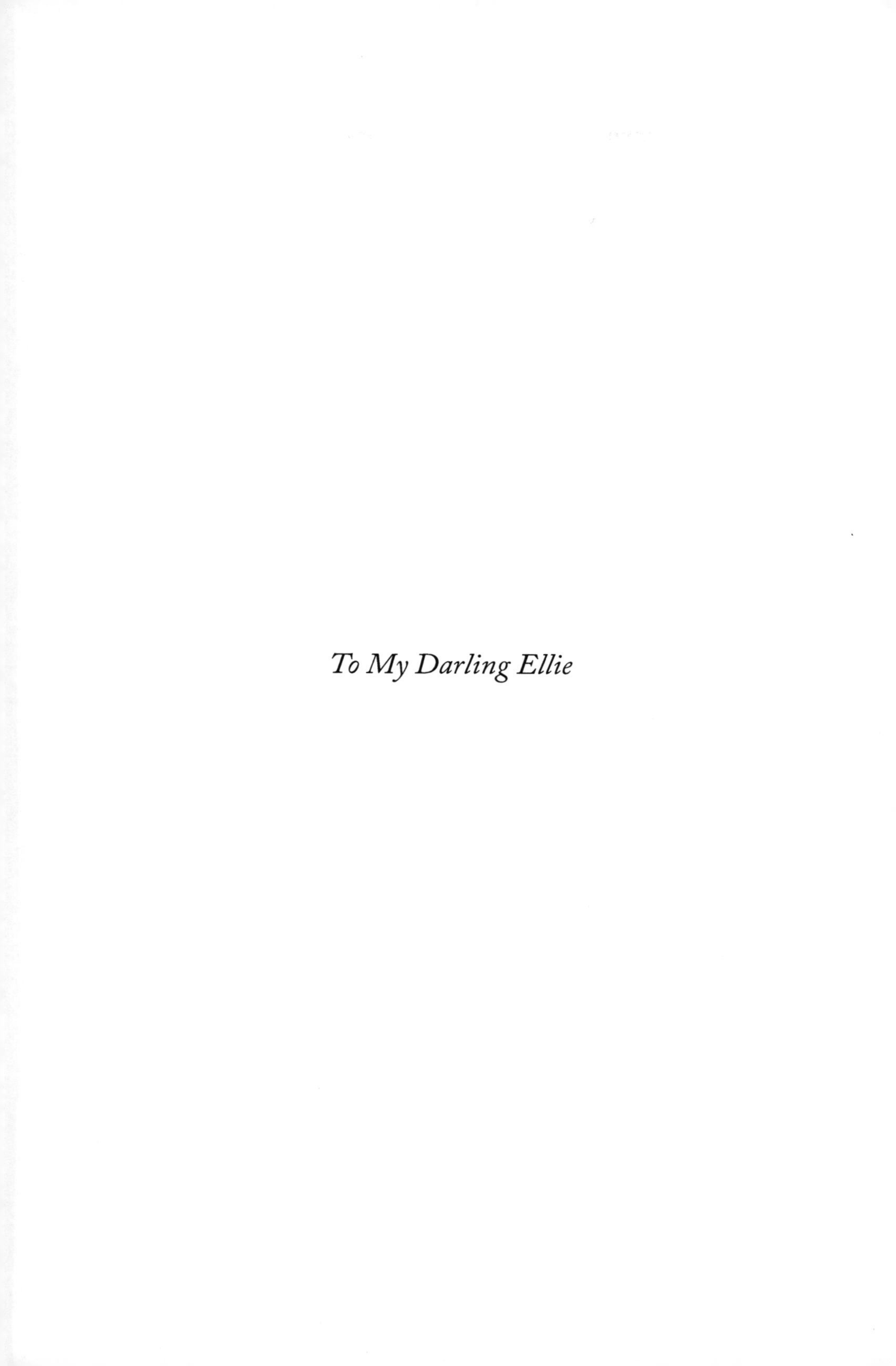

To My Darling Ellie

I'm not a dog, I'm a Lurcher

Attributed to Robbie and Ellie

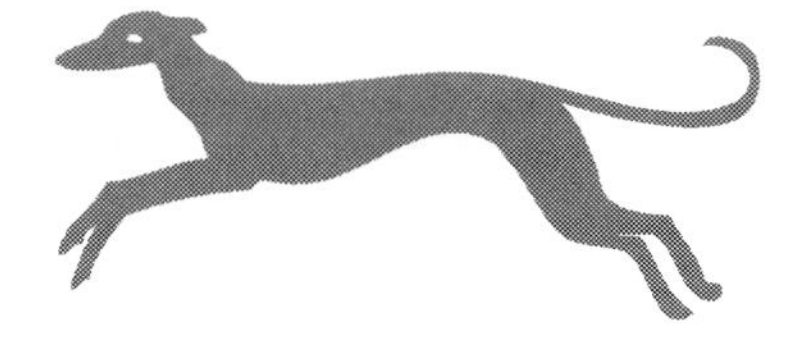

PREFACE

They said you couldn't do it (so we did).

My life has been plagued by people telling me or my animals:

"You can't do it", or "It won't work", or "It will never succeed", or "You won't get in". If I were to have listened to them I would never have got a degree, worked in the City, or written this book.

Ellie and Robbie are a very good case in point: don't get rescue dogs, they are untrainable and likely to be a liability. Music to this girl's ears: as the unconscious mind does not recognise a negative, this meant to me that they are, and were, trainable.

My inspiration for this little literary nonsense offering is years of seeing bestseller lists at Christmas, such as "The Cat's Little Book of Kittens", "The Lady Gardener's Book of Useless Tips", "The Dustman's Guide to Bins" and "The So and So Book of Nonsense".

The sort of books that always sell well in December, to people who have little imagination as to what to get as a present for people who have a great deal of imagination. These books litter our bookshelves, gathering dust after an initial and somewhat bored perusal and being tossed aside in favour of the television, or even more often the iPad or computer game.

I could go on with this little diatribe but I promise you I am coming to a point: why are you reading this book, unless you are at your Great Aunt Dora's, who is dotty about dogs, and there is really nothing else to read? And more pertinently, why have I written it?

I am telling you because every day I am faced with evidence not only of Ellie's mortality but mine. Everything changes, nothing stays the same and there is nothing more poignant than knowing that the dog you love beyond measure is very unlikely to outlive you.

By writing this book, and you reading it, I am attempting in some way to make both Ellie and Robbie, my beloved Bedlington Lurchers, immortal and to share with the world their unique and special characters. This is the story of not one, but two love affairs and the tale of two wonderful animals whom I have been privileged to know.

Now I can finally say "We did it!". Thank you for reading.

Pippa White
Gloucestershire, March 2019.

CONTENTS

AMPHIBIANS, CATS, BATS, RATS AND FEATHERED FRIENDS

ELLIE

You can find full colour versions of the pictures in this book, as well as many others, at:

WWW.BOOKSABOUTDOGS.CO.UK

CHAPTER 1

Google Girl

"Bedlington-Cross, Broken-Coated, Brindle, Lurcher Bitch"

I checked the description for a final time as knowing that the only way to use Google is to input the accurate question if you want a satisfactory answer.

As Google has no regard for grammar I did not worry about the full stop. It was a Monday morning when I started looking for my girl, as my husband particularly wanted a bitch! But all the so called lurchers I had come across on the internet so far looked either not like lurchers or just downright dodgy and certainly not the Bedlington-cross I had in mind.

I decided to refine my search and used the terms above. I paused, checked and then hit the SEARCH button...

She is lying beside me on the bed in the early morning of the closing days of the year, her head on my lap, her paw on the laptop, wishing I would curl up and go back to sleep with her in these days of hibernation. Quite right, too, until later when she wants to race out on the open land under a clear sky and an open field...

The photograph was heart-breaking: the thinnest greyhound-cross lurcher I had ever seen, but I knew she was my girl. I emailed my husband who was away and he simply replied "Ring". I did and everything else is part of this story.

This is the dog who went from rescue to royalty protection in six months, beguiled a duke, kissed a frog and won over countless hearts. She has inspired me to write not only of her but her snowboarding predecessor, other dogs, count my chickens, tell of a rat's parliament, relate of an intrepid cat and to share many other stories of the animals I have had the privilege to know.

But I can tell you one thing right now, despite the fact she rules the household, and takes over three quarters of the bed, is irritatingly, massively, beautiful without needing makeup, demands her rights and has better command of syntax than any dog I know - that Google search changed our lives forever.

CHAPTER 2

THE DECISIVE DOG

"The Girl With The Beautiful Eyes"

We arrived at our destination some hundred and fifty miles, and three hours away, having stopped off at Rutland Water for a quick picnic lunch. We were struck by the suburban nature of this famous beauty spot and, not for the last time, were thankful for living in rural Gloucestershire.

Drawing up at the farmhouse we were met by Jane, Ellie's foster mother, and her two dogs, one dark black excitable lurcher, the other a small terrier. Our girl to be was nowhere to be seen, yet.

Now before we go any further, I must confess to a sleight of name; Ellie's eyes are truly that deep luscious amber hazel brown that charm us all so and embellish so many of these wonderful dogs but I balked at her name.

The "Bedlington-Cross Broken-Coated Brindle Lurcher Bitch" was called Hazelle which somehow didn't feel as if it was going to work for us, and so I had shortened her name in my mind to Ellie even prior to meeting her. This was on the back of Jane saying "You sound perfect and she is already listening to me talking to you on the telephone."

Perhaps, I had pondered: Haze as a shortening, somewhat like Misty or Shadow, common names for lurchers and referring to their slinky shadows and tendency to lurch towards their human companions. But somehow Ellie sprang to mind, being close to but not so far from her birth name.

Her full formal name was bestowed upon her at a later date by my brother in law later in 2009 at Christmas when she was being bumptious.

"Be quiet, Eleonora, and settle down!" he said somewhat irritated by her trying to get his attention.

Eleonora, she became, at times of disobedience and state occasions and it does suit her. It is derived from the Greek for "ray of sunshine" and she truly is like that to me. But I skip ahead...

We proceeded through the porch, into the passageway, where clearly the dogs slept, judging by the number of dog beds, and on out to the long spacious garden with delightful herbaceous borders and an orchard area at the end.

There we found her, apparently somewhat timid, and an incredibly scrawny little thing, even thinner than her photograph - no whiskers or eyelashes, down rather than fur on her hind legs and, it has to be said, not very good teeth.

She was still adorable to me. Anyway, I had already set my heart upon her and to that end endured a lengthy interview by Greyhound Rescue from Stroud to vet my suitability so nothing was going to deter me. I went over to her gently.

"Hi little one, how are you?" I stretched out a hand to stroke her, and though not rude, she was barely the effusive, grateful little mutt

I, probably unfairly, expected her to be. Moreover, despite giving me just about the time of day, she pretty much ignored both my husband, Christopher and my daughter, neither of whom are used to being overlooked. I became very concerned that we were getting off to a bad start. And then to add insult to injury she mooched off inside the house.

Now I consider myself pretty good with dogs, having only come a cropper on two occasions in my life, once in Italy as a child where I was bitten by a farm dog (probably rabid!). And another time, when whilst I was sitting perfectly still on my best friend's frightfully grand grandmother's sofa, her Jack Russell suddenly elevated himself to nose height and attempted to give me some sort of plastic surgery. I have never liked Jack Russells since...

Oh dear, the interview was clearly not going as well as it should! Instructing the others to stay outside in the garden, Jane and I followed the dogs into the house where Hazelle/Ellie proceeded to hide under the kitchen table. Little did I know that this behaviour, far from being timid, was actually a sign of things to come and the first evidence I had of Eleonora's feisty and formidable character.

What to do?

I decided to sit by the table that sheltered our reluctant hound-to-be, pointedly ignoring her, chatting away to Jane in the company of her charming teenage boys who were glued, in that peculiar manner of the male species, to the television and their laptops simultaneously. I do not recall the details of our conversation as I was becoming acutely aware that I was being scrutinised somewhat professionally... I endeavoured to keep my voice low and, hopefully inviting trust, and continued my attitude of studied indifference to this canine perusal. I remember trying to stay relaxed whilst straining to hear any little sound or change of breath or posture which might indicate what was going through her mind.

Half turned away from her, imagine my surprise when a tentative paw landed delicately upon my knee and I shifted to look into the most enchanting and expressive eyes I had ever seen.

"Would you like to go for a walk?" I enquired of this clearly opinionated canine.

She gracefully acquiesced and we two left to meander along a path that adjoined a stream near the house. In the golden August sunlight, Ellie happily trotted along, shuffling and snuffling contently through the early fallen leaves, all shyness apparently lost, and she was positively jaunty. I, though. was somewhat self-conscious due to the disapproving looks I was receiving from the other dog walkers whom we encountered, as, despite the wonderful care and attention she had received, she was not yet anywhere near looking her best and clearly, judging by the judgemental looks of those we encountered, I was to blame!

No matter, we enjoyed ourselves and returned to invite C and M to join us. After a second and more confident stroll, we went back to the house and finished the negotiations for her departure, including a rather larger sum than we had been expecting to pay, which was however rewarded by Ellie's highly unexpected next move.

On leaving the house, off the lead, she marched straight up to OUR car and stood by the boot. Clearly, her mind was made up, she had vetted and approved us and now was going home. It was a sign of her very strong character and things to come. She is the only dog I have ever lived with who decides when the walk ends and it is time to go home, for instance!

That evening was a great success, Ellie is a most curious animal, if she can see over a wall or around an obstacle she will. She inspected the pond, ate a good supper and enjoyed loads of cuddles. It soon became clear that all the good and much-discussed intentions of the dog sleeping downstairs were to be abandoned when she proceeded to lead the way to bed. Consequently, C and I spent a most uncomfortable night with a somnolent, immovable and very dense little lurcher at the diagonal between us. Now she has the best of sleeping arrangements, but that is another story, for now, she was home and making herself truly the centre of the household and attention. Nothing has really changed ever since!

CHAPTER 3

THE CLOAK OF INVISIBILTY

The Previous Day

In the week between the Google search and going to collect her, I had spent many hours reminiscing about my previous love; I had been broken-hearted after the loss of my beloved Robbie, a strawberry blonde Bedlington lurcher, with more than a touch of Saluki and a temperament to match. It had taken five long years and a bit of a journey to reach the point where I was prepared to test my heart again for another dog.

Meanwhile, I had also lost Ferdie, a deerhound-cross lurcher of the sweetest and most dumb nature I have ever come across. She died in the sunshine, at home, with the cat purring beside her, and eating chocolate (Green and Black's) for the first and last time in her life when the vet came to put her to sleep at home as she had incurable cancer. Luckily she had not yet suffered too much pain to my knowledge.

Robbie was not so lucky – my best friend, companion, and shadow of nearly 14 years had had a stroke, which came without warning, and I woke one morning to find him gone from my bed. I knew this only meant one thing: Robs was dying or dead. Ferdie had taken his place under the dressing table, except I found Robbie still alive behind the sofa in the sitting room and called the vet, thereby subjecting my beloved to two days of agony: separation from home and all the ailments of a stroke without my stroke to help him. He died in the car, thinking he was going home. He was one of the most special people I have ever met in my entire life: loyal, loving, funny, speedy, clever, beautiful and difficult – he was going to be a hard act to follow.

So, the day before we collected Ellie had been exciting and sad by turns; a happy adieu to my dearest friend after a serious period of mourning, in anticipation of a new relationship.

I had long contemplated what best to do with the ashes of Robbie and Ferdie, which sat in the house in their ornate carved wooden boxes, as I had been unsure where best to make their final resting place, or should I say hunting-ground. I was reluctant to bury them in my garden as I did not know how long I would live there (I am still here) and the family home had long been sold, so I had bided my time until a solution became clear.

Suddenly it did – their ashes were carried in state to a favourite spot of theirs where I so often had taken them to walk, and where the view is back from Wales across the Severn Estuary to where we live now in Gloucestershire. My daughter and I had nicknamed it "The Top of the World Place" when we lived there. It is indeed a most rugged and atmospheric piece of country, not unlike the Yorkshire Dales, with broken lichened stone walls and wind-carved trees stunted by the weather and somewhat bleak but utterly compelling. It became, for us three, Robbie, Ferdie and me, a place of succour and new hope in turbulent times when we lived nearby at Rose Cottage.

I decided that we take their ashes there to release them to the wind and the river and the sky at the place where the dogs had been so happy. Not only was this a favourite spot of ours, but somehow the long approach seemed a fitting last walk, along the flat forest path flanked by ferns, and varying coniferous and deciduous plantations, until the right turn up the hill through an archway of ancient beech and a sloping upward hike to turn left at the top, where the view changes to upland pasture. Then, suddenly by a little mound, it opens up to encompass the huge space of the fields, forest, estuary and sky. A place for dreaming, and adventure, and resting, and picnics. From my study here the other side of the Severn River I look west back to that same hillside. Somehow there is enormous reassurance in this – that somewhere out there their souls are still enjoying the wonderful pleasures of that upland world and I can still see and connect to those hills, and by inference, them.

After a jolly good picnic, I unpacked the two white bags of ashes; there is a picture of me standing with one of the bags in my hand, looking somewhat suspicious with this packet of white powder. Then M and I went to the wall that is "The Top of the World" as we had done so many times when she was little and there we returned the dogs to the earth, the wind and the sky. The ashes flung out and scattered against the air and then suddenly were there no more. It was almost as if they had vanished in a puff of smoke as if by magic. It was spellbinding and it felt a fitting adieu and a farewell to those two beloved companions who I still love so dearly and cherish their memories.

Writing this now, I hear once more the early warning first squawks of the dawn trumpeter and the familiar grumbles of the other sleepy birds and I contemplate that this little committal and memorial service meant our new friend would have a free run and would not be living in their shadows, as they had been returned to the earth and were running free forever. I can see up on the hill, outside my window, the dark grey of night is lifting, the stars have retired, and the dawn light betrays that the sky is clear this day in late October.

We are delighting in the most glorious autumn I ever remember – sunny warm days, vibrant berries of every variety tumbling in abundance, roses still in blousy bloom, all colours of pink, pale peach, crimson and creamy white, now punctuated by rosehips. The hedges adorned with fluffy Old Man's Beard, and the fields, now shorn of their summer crops, ploughed and in quiet anticipation of new yields. The symphony of the autumn birds, the crows and wheeling seagulls, the tea-leaf swarming of the starlings, who once again appropriate the church roof after the departure of their aristocratic cousins, the swallows, to warmer lands.

The orange flare of the sunset merges into the deeper maroon of the oncoming night sky and a happy dog lies before the fire, now lit to accompany the cooler evenings as the light draws in. There are long cosy nights in bed, dreaming of the place at the "Top of the World" where lurchers go racing, racing, coursing the stars and dreaming at night once more.

ROBBIE

CHAPTER 4

My First Love

Robbie

Where do I start in the love song to my beloved broken coated strawberry blonde Bedlington cross legend?

Early in my first marriage, I started to yearn for a dog, particularly a lurcher as I had so loved my mother's lurcher, Benjie, whom I had known since my late teens. I now wanted a companion of my own, so I began to scour the local paper for rescue dogs. One day when my then husband came home, I showed him the picture of a lurcher bitch being advertised by a local rescue centre.

He paused and looked at me and said, "What is wrong with this one?" pointing to a picture that somehow had completely bypassed my awareness.

There was the most appealing pair of eyes with just a hint of mischief, dancing off the page of the rather pathetic appeal for homes for some 12 dogs. I nodded and said I would ring the next day. I still have the newspaper cutting somewhere!

Robbie was being housed at a rescue centre down the hill near the river, so very close to us in St.Briavels in the Forest of Dean, and immediately after our call, we went down there to see him. His was a heart-breaking story: he had already been rehomed once but had been "returned" on the basis that he was too "skinny"! In my book people who make that call about a young dog, nine months roughly and more precisely a lurcher, who has drawn a pretty rotten lot in life so far, do not deserve to own a dog at all.

Robbie was brought out of the kennels: skinny, yes, bouncy, very, blonde, extremely, appealing, beyond measure. Just a puppy still, all legs and fluffy down; he immediately jumped up to me and I fell in love just like that. Immediately and forever.

Attempting to be sensible, I said to the kennels that we would consider him and I went home and called my mother. She had been extremely dismissive about acquiring a rescue dog as we always owned our animals from eight weeks and she saw only trouble ahead. She did though agree to come and see him at the kennels with me that afternoon and so Robbie was brought out for the second time…

"Oh, he's just a puppy!" she exclaimed, her heart melting in the face of the sweetest charm offensive I have ever witnessed in a dog. Now lurchers generally, but Bedlington crosses in particular, either switch on the charm as if they are the sole representative for rent-a-dog and "you can't ignore me at all" or "you can't leave me this way" eyes and so on; OR they totally and utterly blank you, depending on their opinion, swiftly formed, of your character, and furthermore, your possibilities. For instance, in Robbie's case vets were to be firmly shunned, (and he had enough experience of them as you will soon see), but Ellie loves the vet! Though that might be something to do with the dog biscuits and attention. So it is all a matter of opinion and the one thing that lurchers are is extremely opinionated…

Well Robs certainly turned on the appeal that afternoon and it was rapidly agreed that, no he couldn't be left at the kennels, yes, he needed a loving home urgently and, definitely, that home was with me.

I returned the next morning (interestingly I needed no vetting as I did with later rescues) and at reception announced that I had come to collect Robbie. Then we suddenly heard the most extraordinary sound somewhere between barking, singing, calling and whimpering coming from inside the kennels.

"He's heard your voice and he knows that he is going home," said the girl on reception.

And that was it – I knew and he knew that he was going home with me for nearly fourteen years. A love affair that only intensified with age and experience, an enormous privilege of the kind that makes life worth living, a love that endured numerous accidents, naughtiness, the birth of a rival, (my daughter), divorce, six house moves, town and country, just to list a few of the major life events we shared. There is so much to tell you and so many adventures to recount and relive, such happiness to share.

Anyway, Robbie was violently car sick all the way up the hill on the way home, already in love, I was worried that this did not bode well, as our lives living out in the sticks involved a great deal of car travel. My fears were to prove unfounded as Robbie became a most determined and willing car passenger in the ensuing years.

CHAPTER 5

Home For The First Time

We lived at that time in a former Coach House on an estate near St.Briavels in the Forest of Dean. It was one of the happiest times of my life, set in the magical surroundings of this enchanted place where the open fields at the top of the hill overlooked the Severn Estuary from the East. I coveted the mysterious dilapidated wall garden with its old decrepit glasshouses which I yearned to do up, and beyond it, explored the ancient woodlands with all sorts of exciting flora and fauna. Wild lilies and orchids grew there, lianas of wild clematis and honeysuckle, delicate white wood anemones and bluebells, and a stream, more like a little river, ran through. It was to this entrancing place, full of magic and potential, that we brought Robbie home that day.

On arrival at home Robbie jumped confidently out of the car and preceded me to the house. After making himself at home on the sofa we had our first little contretemps. Needing to do a wee, he tried to go in the corner of the sitting room! I began to panic slightly but immediately took him outside and explained that there was the correct place. Bless him, he had been in the kennels so long that he had unlearnt any previous house training, but he was a fast learner. He never tried that again and always went outside even when his leg was in plaster!

The second little difference of opinion was an indication of Rob's intelligence and cunning and a sharp learning curve for our new friend. The Coach House was a converted stable and some of the loose box divisions had been left to divide the sitting room from the stairway and hall and the kitchen beyond. Around five o'clock on that first day, Robs began to get hungry again and took himself into the kitchen to investigate the culinary possibilities. Through the iron railings on the upper half of the loose box, I spied him, paws up on the kitchen surfaces, scouting for food. Taking advantage of the cover provided by the loose box division I, possibly somewhat sneakily, chastised him remotely, or so it seemed to him, as he couldn't see me, as I was not in the kitchen, I must be a disembodied voice!

"Robbie, no! Down, bad boy!" He looked round startled and though he could locate the source of the voice he couldn't see me. The deception worked, those paws never met the kitchen surfaces again.

They say lurchers are born thieves. Well, that's as maybe, but my experience is exactly the opposite: Ellie and Robbie have been exemplary in their attitude towards taking food and have shown remarkably little tendency towards dishonesty.

Once Robs took a chicken wing from the table when he was still quite little. He hid it under his shoulder and lay in his basket under the kitchen table, looking guilty as hell, making no attempt to eat it and even less attempt to prevent me taking it from him after a little scolding. He never stole in the house again, though there was once a problem outside once, but that is another story.

Very soon, Robbie proved himself to be clever, loyal and extremely affectionate. But the true mettle of his fantastic character and his many talents had yet to be revealed.

CHAPTER 6

Home Alone And Trashing The Place

Robbie proved to be a fast learner and settled in with us very quickly. However, he suffered from separation anxiety, which was not surprising given his previous experience and became very alert to clues that I might be going out. Heaven help him if I had to leave him behind, as was occasionally necessary. He was so attenuated that even when I applied mascara or makeup or even THOUGHT about going out, he would start campaigning to be with me, as does Ellie who appears to share this uncanny canine sense.

Although Robs did not take kindly to being left at home alone, it was necessary, sometimes if I was going shopping, and it was too hot in the car or too cold or the venue was not dog-friendly. However, he devised the perfect way to demonstrate his ire at the situation. On arrival home I would regularly find all the cushions from the sofa and chairs in the sitting room strewn all over the floor and the place looking as if he had invited all the local dogs over for a cushion bashing party. Naturally I assumed that this behaviour occurred sometime after I left as a result of rising anxiety at my absence, but once again was proved wrong.

On one occasion, I was going to John Lewis and I had forgotten the measurements for something and went back to the house to collect them. A mere 30 seconds after my departure I opened the front door to find Robbie in glorious flagrante with the furnishings. He had already redistributed at least three of the tapestry cushions which decorated the sofa and was surprised to see me especially as his mouth was full with the fourth! Needless to say this interruption did not deter him from future furniture rearrangements, he continued to make his feelings clear in this irritating, but fundamentally harmless way, as he was never a destructive dog at all except for killing the odd soft toy.

He was funny about soft toys later in life: M by age four had acquired a pretty large collection of small stuffed toy animals, the result of being thoroughly spoilt by her Grandparents and me, but she was sometimes away with her father. On these occasions, Robbie would raid her room with great glee and, with an air of great triumph and stealthy acquisition, he would bring downstairs whichever poor bear, or cat, or seal, or dog, or other unsuspecting toy had been abandoned upon the bedroom floor.

These were never harmed, just presented to me, as if to say, I could but I shan't, kill them. Robbie's own toys did not survive so well and were regularly relegated to soft toy graveyard or Heaven, whichever way you want to look at it.

CHAPTER 7

Accidents Galore

Life was never dull with Robbie, after we brought him home on Valentine's day 1992. My husband never failed to point out that he couldn't take me out to dinner that day as it was the dog's anniversary and I would whinge all evening. This was not entirely unfair as we once went on holiday to Madeira and I missed Robbie so much I spent most of the time talking to a spaniel who lived near the hotel! Needless to say we did not go away again.

Anyway life was never dull but sometimes tedious, especially during our first summer and early autumn together, when the weather was awful. It rained nearly every day and the ground was pretty treacherous and slippery. One day Robbie and I were out in the woods and he was running at full pelt as was his wont but unfortunately he ran at high speed into me with a terrible yelp. On examination it became obvious that he had done something serious to his front right paw.

I had to carry him some distance back to the house as he was unable to put any weight on his leg. I examined his very muddy paw but he seemed in excellent spirits, so I held my breath and hoped for the best although I suspected a fracture. Unfortunately I had some important clients coming to buy some flowers and was unable to contact them to cancel, so I had to wait to get him to the vet. No matter, he seemed cheerful enough…

Eager as usual to ingratiate himself with any visitor he appeared his normal happy and extrovert self when my clients arrived. However, after they had left he appeared to be in great pain, so a visit to the vet was imperative.

The damage turned out to be a compound fracture of the toe which needed resetting and an overnight stay, which was bad enough but worse was to come. Not only did I have to drive in the dark, a howling gale and pouring rain an hour and a half to the nearest veterinary hospital where he had to stay overnight but the next night after I had collected my poor dog with his newly plastered foot, our house was flooded throughout the whole of the ground floor.

This was a major disaster for a number of reasons: 1. My husband was away in Thailand with our suppliers for our parchment flower business 2. The coir carpet was completely wrecked 3. The new brochures for our parchment flower business, our first brochures in fact, were largely destroyed by flood water, though we managed to salvage some. 4. And most importantly, I had to take Robbie post-haste back to the vet as his plaster cast which was to be on for six weeks at the very least was not allowed under any circumstances to get wet as this could lead to his foot rotting.

Much to my chagrin and Robbie's fury, he was duly deposited once again at the veterinary hospital and I returned home to bail out our house which was now bathed in the most glorious late autumn sunshine after the terrible weather the night before. I felt very alone until the house was sufficiently dry to house my temporarily disabled pet and he could return home.

But these dramas were not over, needing to cover the plaster cast every time he went out so it was waterproof was tedious, and also in our freezing house, keeping the foot warm was not easy. I resorted to the tiniest baby socks I could get from Ladybird in white, blue and red! My enduring memory of Christmas 1992 was of Robbie at my sister-in-law's house with her newly born twins and Robbie sporting a very festive pair of red socks over his somewhat grubby white plaster, looking for all the world, like some reluctant canine Santa Claws.

He was grounded, unexercised, miserable and grumpy but bore the hardship well for six weeks or so until one day he snapped at me when I was putting on the plastic bag and elastic band to protect his foot when he went outside. Realising something was very wrong indeed, as this was entirely out of character, we returned to the vet where I am afraid I declared that if the toe was not mended may be more drastic action was needed.

It wasn't. Once the cast had been removed, and the cause of Robbie's bad temper revealed (the cast had been rubbing against his spindly legs causing bloody sores), he was declared healed. This was such a relief, as I had feared that due to his unfortunate start in life and his extreme thinness when we got him, he might not have had very strong bones but it appeared that my concern was unfounded. Two weeks later, after an interim time of extremely boring walks on the lead, he was once again, running at full speed!

We had many accidents over the years: dashing straight into a barbed wire fence, so a horrendously torn ear which spouted blood like a fountain all over my car on the way to the vet and became infected on a bank holiday, of course. (The ability of dogs and children to develop illness or have accidents on Christmas Day or some other highly inconvenient time never ceases to amaze me!) He sustained a severe and deep cut in Dumfries, necessitating a very bloody drive back home, and we endured some very severe looks from people at the motorway services at the clearly desperate poor mistreated dog. At least a dozen dew claw accidents until they were removed...in all 13 accidents, never mind a major illness aged 9 which nearly killed him, but meanwhile back to full pelt, his favourite pace outside the house....

CHAPTER 8

Speed And Snow

Robs was definitely the fastest dog I have ever come across and one of the highest jumpers, although I have seen higher leaps by lurchers at the Game Fair. His cornering was exceptional and consisted more of a three quarters airborne pirouette than a grounded turn. Not only that, despite the early fractures, after a couple of years of a decent diet he seemed to be very strong and possessed the combination of speed and doggedness which could fell the chunkiest of men.

One early summer, we had some friends to lunch and decided to take the dogs for a post prandial stroll. The grass was lush and just high enough to obscure Robbie as he was dashing around showing off. He was madly exhibiting that peculiar lurcher mannerism of rushing straight at someone at top speed and then veering off at the last minute so as not run into them. At least that was the theory.

That day he was doing this magnificently until he made not just one, but two major miscalculations in the space of a few minutes. Two of our guests were well built men, to say the least, weighing more than 30 stone between them, their very stature convinced me that there would not be a problem with Robbie knocking them over, as had happened to me in the past.

Well I could not have been more wrong: Robbie was below grass level and at full speed when he impacted with my friend Nick who went down like a nine-pin to shrieks of laughter and some consternation on my part. David and Goliath came to mind when witnessing an 18 kilo dog felling a man approaching 90 kilos. Speed clearly adds weight!

Worse was to come, with Nick dispatched, next Felix, who was even better built, toppled over in the grass by the giant seeking canine missile. Somewhat concerned by this further development, I called a halt to the walk and we returned home, Robbie by turns embarrassed and thrilled with his unexpected conquests.

He was though very stiff the next day.

His athleticism never ceased to amaze me. One snowy day, one of the first he experienced, we went out early into the enchanted world of virgin snow with the hoar frost gracing the trees like sugar icing and tinkling like as it cracked and broke away from the branches. Our breath clouded in the freezing air and feet and paws were gliding around on the slippery ice beneath our feet. Going across the little bridge over the stream was quite hairy until we reached the favourite field.

This field had the perfect slope for a bit of playing and Robs raced up through the snow, the fresh powder spraying all around him, so he was almost hidden from sight until he reached the top where he turned against the rising sun and the powder turned to golden light around him against the horizon. Snaking up into the air and turning as he jumped he was as awesome a sight as any Olympic gymnast. How he played that day, twisting, jumping, turning until he seemed to transform into a flying dog flung out against the gleaming sky laden with that extraordinary rose gold hue that characterises the snow laden heavens.

Several hours later, one totally exhausted and very happy dog and I thawed ourselves out by the wood burning stove for the rest of the day. Thereafter whenever I told Robs it had snowed he would rush downstairs, eagerly anticipating the rush of cold and singular joy of snowboarding on the white blanket that would stretch as far as the eye could see. The snow times were some of the best times, even in the moonlight after the snow had finally finished falling when we would take a midnight walk, bewitched by the fairy tale quality of the sparkling crystals and the dripping ice. One time the freeze had come in so hard and fast after the preceding rain that a group of hawthorn up the hill sounded like cracking glass when their branches moved in the wind, so encased by ice were they.

When Ellie first saw snow she was curious but wary. After venturing out gingerly, she returned back to me very fast with a disgusted look on her face, convinced that I had engineered this mesmerising cold which was hurting her paws. She has since sort of come to terms with snow and ice but still prefers the seaside where she loves her midnight moonlit walks by the murmuring silver sea.

CHAPTER 9

Deceit And Five Bar Gates

Robbie was a bit sneaky about talents that a Houdini hound might find useful, just in case they were ever required for the great escape. I sometimes used to think that much of the time Robbie appeared to be dreaming, he was actually fantasising about his next elusive adventure. I do not think this arose from lack of love or security for he was devoted to me, but just pure devilment and a sense of inquiry and adventure, not dissimilar to the great explorers. Ellie is a homebody who becomes anxious when we are out of sight, Robbie loved nothing better than a jolly good adventure so long as there was a good dinner, and a warm fire, and a comfortable bed at the end of the day.

And so to the five bar gate that Robbie had told me that it was impossible for him to jump. He had made it abundantly clear that he was incapable of hopping over anything higher than the merest low tree trunk but I sometimes wondered whether he was being entirely truthful. If we reconvened, having parted company some small distance back, I would ponder where and what he might have traversed to return...

And one glorious spring day my suspicions proved to be well founded. On the way back from our early morning walk, I often walked, with permission, through our neighbour's garden after he had gone to work. It had a little brook running through and D had erected a sweet little bridge over this. We often worked together on our growing gardens, swapping plants and ideas as the seasons went by. D grew many wild flowers from seed collected from the woods where he walked each weekend and the brook was fringed with water loving foliage and attended by many pond loving creatures. It was a place of great peace and unsophisticated beauty.

Robbie was not so keen on it though and, whilst I took the left fork through the garden and over the stream, he would take the right fork joining me back on the drive some 200 yards from the house. I never really gave too much thought as to how he exited the garden.

Robs had always been adamant that he could not jump anything, always preferring to go under any fence or gate in that odd snaky lurcher fashion of head first, and undulate the long back into a concave curve, and then slither under the gap…except he was not being entirely straightforward as I discovered one day when my suspicions were confirmed.

I came round the corner from the side of the house to discover Robbie arcing in full flight over the five bar gate at the apex of the jump, full stretch and full speed ahead, that is until he realised that I had seen him and there was just the slightest hint of contraction as he registered oh **** I have been rumbled.

It flashed through my mind that he might in be danger of executing the most spectacular belly flop onto the gate so I averted my gaze and kept very quiet as he resumed his concentration and powered over the gate somehow accelerating in mid-air and landed the other side. He strolled over to me with a cheeky grin, which hid some embarrassment,

I think.

"I thought you couldn't jump ANYTHING."

Robbie gave the lurcher equivalent of a Gallic shoulder shrug and slunk rather jollily back home where he ate a hearty breakfast and fell into the wary slumber that characterised the waking hours. One eye ready to open just in case another walk or a car ride was in the offing, or to react fast were I to commit the cardinal sin of planning to go out, go anywhere, without him. He was indeed my true shadow, unless he was exhibiting his high jump prowess over gates, or streams, or fences that he could NOT, NEVER, EVER jump at all!

CHAPTER 10

Dead Lamb And Locked Up

However, this exceptional speed and agility was of no use for coursing as he showed absolutely no interest in killing anything, not rabbits, nor deer, nor sheep. Though I did once have a problem with a lamb which was not entirely Robbie's fault.

Out walking one day, we came across a dead lamb which, I suspect, had been savaged by a fox (his arch enemies). Robbie, being young and inexperienced around livestock, became very excited indeed. Oh no, I thought, trouble ahead and there was indeed. Despite the fact he had never shown any inclination to chase sheep, he attached himself to the dead carcass and would not let me near him at all to take it away. I became increasingly concerned at this atavistic behaviour and totally uncharacteristic growling, especially as I was well aware that the farmer would be entitled to shoot him on the spot if he came across us.

Needless to say it was pouring with rain, as usual, and I mean really pouring with rain and I resigned myself to a long wait out in the atrocious conditions, having retrieved some more waterproof clothing at the risk of leaving him for five minutes. After five hours out in the relentless rain watching the dog guard this dead creature, without making any move to eat it or otherwise deal with him, both Robbie and I were more than soaked to the skin. I was very cold, angry and frustrated having shouted at him periodically to get him to leave the poor little thing alone.

Eventually he gave up and leaving the wee lamb, which looked more like a wet woolly mat by this point, he wove wearily back to me and conceded defeat. I was furious and, more, I was determined that this behaviour which would endanger his life and, potentially that of others, was never to be repeated.

I beat him (the only time I ever did) and shut him up in one of our outhouses in the dark for the night. I spent a sleepless half of the night and finally gave in at two in the morning, venturing the 300 yards in the pitch dark, and still sodden night, to retrieve him. He was most effusive and grateful in his affection and full of promises to be good when we got back to the warm and dry house.

He never looked a sheep in the eye again and indeed, thereafter, his circumvention of any flock bordered on the comical.

CHAPTER 11

Houdini Hound

Robbie was very honest about food and never stole any after the chicken wing but he was less honest about escaping and running off. He was the ultimate escapee. Early on I noticed that he had the most peculiar habit of disappearing at weekends. The only thing different about weekends was that my husband was at home, rather than out selling our parchment flowers which he did during the week.

Friday afternoon or Saturday morning, when Robbie would duly take to the fields and hills, I would spend many an hour searching and calling for him. I consider it very dangerous to have a dog on the loose, even in a protected area such as the few hundred acres in which The Coach House stood.

Initially he would just dash out of the front door which was our main door, but as we grew more vigilant, he grew stealthier. Oddly enough, he would always return for dinner and bed (with me), even if somewhat dishevelled. However, the plot thickened when we put a stop to the front door dash, put up a fence and gate in the garden which was originally open to the fields behind the house, and Robs was still managing to get out and go off to the open road! Every inch a Gypsy dog.

Which brings me to his origins. We do not know where Robbie came from, other than he was found in a local town called Coleford. He certainly came to us reasonably well trained, with a penchant for fast food, especially fish and chips and curry, and a real fear of men. Sometime after he came to live with us I found a really deep dent in his skull and, fearing he might be affected by this, I rushed him to the local vet who, after examination, stated it would be too dangerous to adjust and we surmised that it might have been effected by the end of a broomstick. As he did not seem any the worse for wear, except for his occasional fearful behaviour, which he never exhibited around me, we left it. But I always wondered what had happened to him before.

Oddly enough, he seemed, after settling in, to go through a second puppyhood, being naughty and tiresome and clingy in the way of the very young. The homeopathic remedy, Pulsitilla, sorted him out but he needed it from time to time for the rest of his life as he never entirely got over his separation anxiety, unless he had decided to take himself off adventuring, rather than being left alone.

Anyway he turned out not only to like curry but to be the most insufferable snob so we would often joke that he was really Lord Robert of Aylesmore or in a former life a Maharajah who had gone to Eton. Whenever we went to people who had far grander houses than us, Robbie would always insist on entering through the front door – the grander the house the more insistent the dog! I once read an article in Country Life magazine by a writer who had a lurcher called Frances, who displayed a similar trait of refusing to use the back door! Lurchers certainly know their rights and have no hesitation in demanding them. Ellie, possibly, even more so than Robbie.

Having put up a fence and gate and guarded the front door with vigilance, we then discovered a number of lurcher "holes" under the hedge to the right of the top of our garden. Some of the holes were so deep I would ask Robs if he needed his passport as he was clearly planning to go to Australia! Anyway, for some months, he was escaping through a little copse and down the bank from the terrace outside the dining room and away to the hills again.

So it was time to trellis the terrace and stop up the gaps in the hedge…we would crack the code of the great escape artist, or so we thought….

CHAPTER 12

THE GREAT ESCAPE

To the left of the trellised terrace was a steep bank which led down from the garden, and to the right, there were steps from the upper storey of the house which were at least twelve and a half foot high, the same height as our ceilings. So once we had erected the trellis and painted it green, we thought "That will teach him!" Actually this turned out to be a turning point in my life as whilst painting the enclosure, I suddenly realised that I was pregnant. This was something of a shock and certainly would change our lives irrevocably.

Our fencing efforts were all in vain, as some weeks later it became apparent that Robs was still on a mission, but by now, I was beginning to wonder whether it was truly wanderlust or just pure devilment. You see Robbie would elude us yet again, but remain outside for only a short time, until coming round and tapping cheekily at the front door asking to be let into the house.

My husband had had enough, "I am going to stake him out". He announced one evening with a glass of whisky in hand.

I went to bed, as I was disinclined to sit outside in the dark, particularly as I couldn't partake of a wee dram at the time, being pregnant, and left them to it.

He waited some time to catch Robs out and our lurcher was at first way too wily to give away the secret of his path to freedom, and so my husband retired inside to the dining room. Leaving the outside light on, he continued to covertly observe the dog and possibly enjoyed another wee nip.

The next morning R got up early, which was unusual. The dog and I were always early risers as we liked to catch the dawn on our walk through the estate, but my husband preferred to get up later. Over breakfast he told me what he had seen.

"You will not believe what I saw last night and I actually thought I might be seeing things. Robbie is getting out at the top of the first floor steps!"

Apparently he had watched Robbie check the coast was clear and then literally helicopter up to the top step which must have been some twelve and a half feet high, walk down the steps and away to freedom. The sensible thing was not jumping down the other side as using the steps as the jump down might have been injurious.

Well, we knew he was bright but really! What we didn't know was that lurchers, and in particular, Bedlington crosses are renowned for their elevator prowess. Some years later I watched in amazement at the Game Fair at Ragley Hall in Warwickshire one lurcher after another compete to jump the highest obstacle which turned out to be around 14 and a half foot which put Robbie's best into perspective but he was untrained, of course.

Ellie does not like jumping heights very much but much prefers the long jump particularly if it involves as much water and mud as possible preferably sprayed all over the nearest available family member.

After being trounced Robbie decided to retire from the Great Escape Game and soon found other diversions which continued to amuse and infuriate us by turns. There was rarely a dull moment except when in recovery from the latest mishap.

CHAPTER 13

Locked Out

Robbie had something of a comeuppance, one of the very few times we were away, and his sly elusiveness nearly became his mistake.

We had asked one of our neighbours to look after him whilst we went to Madeira for a week. He had a collie, who Robbie really liked, so it seemed an ideal arrangement. However, after some debate we decided that he should sleep at home at night, as he preferred to sleep on my bed with all, I presume, the associated smells, to say nothing of the comfort.

One evening our neighbour duly locked up our house, assuming Robbie was in as he had called him in earlier, and presumed him safe. But, little did he know, the dog had snuck out quietly whilst the door was still open.

So Robbie found himself locked out at night in the pitch black, a situation which was entirely alien to him, at least since the lamb incident, and highly unsatisfactory. He must have wondered what he should do. In retrospect, I was very glad that we had not heard about this incident whilst we were away, as being very stubborn and overprotective, I would have insisted on our instant return.

However, Robs was even more resourceful than we had previously appreciated. Clearly worried by this unforeseen development of nocturnal abandonment, he did something eminently sensible.

He went round to our next door neighbours who lived the other side of the coach house, which we were in later years to acquire, and commenced upon the canine equivalent of banging on the door, whining and generally making sure that he was heard.

Luckily, it was a fairly calm night as it was rare not to have a howling gale at Aylesmore which would mask any outside noises. Our neighbours suddenly became aware of an unusual sound outside the door. They went to investigate and found a very worried looking lurcher, who made it quite clear that he needed shelter for the night, and without more ado walked past them into the house.

The self-imposed visitor did not stop there. The neighbours owned two cats, but luckily Robbie was trained NOT TO KILL friendly cats, so he wisely ignored them and settled down for the rest of the evening with the neighbours and associated animals. When it came to bedtime, though, there was a problem: R and N usually had the cats sleeping with them in their room and Robbie made it quite clear that he did not sleep downstairs, he slept upstairs, in bed, and he did NOT sleep alone.

The upshot was that the husband and Robbie spent the night in the spare room and the cats and the wife slept in the other bedroom.

In the morning Robbie was fed with cat food which apparently he most enjoyed and then was "rescued" by the other neighbour, who had been mortified on going to our quarters to find them empty and completely devoid of any lurcher. Robbie thanked R and N for their hospitality with a lick, and treated them with great affection, and always remained a respectful distance from their cats thereafter. D told us this story with some rueful looks and a lot of embarrassment and I am afraid that, unless I had to be away for work, I refused to leave Robbie again. This was despite the fact he had clearly demonstrated that he could manage very well on his own and was extremely resourceful.

CHAPTER 14

Biscuits And Parcelforce

We are frequently told that dogs are greedy and will eat anything and my father was firmly of this opinion. But neither Benjie nor Robbie nor Ellie were or are greedy in the same way as other dogs. Robbie was, and Ellie is, inclined to leave their dinner if: a) not hungry b) don't like it c) they are cross about something. Ellie is greedy about cheese and pizza, Robbie liked chips, Ellie eats raw bones, Robbie hated cake and biscuits or anything sweet, Ellie loves chocolate – so one cannot make sweeping statements or draw universal conclusions about their culinary preferences!

All my dogs have exhibited quite precise and clear predilections for different types of food – as I already mentioned Ferdie died in the sun eating chocolate for the first time. I occasionally give Ellie just the tiniest morsel as I know it is not good for them to have large quantities. But what really gets her going is cheese! Recently after a spate of buying the most revolting processed dog treats from Tesco I have decided to make Ellie homemade snacks.

We started with off cuts of bones from our wonderful local farm shop, and then I decided that I would come up with some healthy recipes for dog biscuits, which are not sweet, are gluten free and are packed with nutrients. This came about also because I had made some parmesan cheese biscuits for a party, which were a great hit with Christopher and Ellie, and we also have a great source of dripping from the farm shop.

Ellie also adores Goji Berries, which the wolves eat in Siberia, and even more recently, she has acquired a liking for dates. The diet of camels and the salukis. It is curious how even carnivorous animals will naturally supplement their diet with fruit or vegetables, particularly fruit for the higher sugar content.

Saffie, the golden cocker spaniel of my youth who features little in this tale, would cruise along the bottom of the raspberry canes when the berries were ripe and so we always knew not to bother to look below a certain level. In our Vineyard, the old fable of the Fox and the Grapes became a daily enactment by the foxes, when the vines were just ready for picking! It is amazing how often pub signs often draw on old country lore for their inspiration such as the Fox and the Grapes, The White Hart, The Fleece Inn, each an Aesop's fable in their literary breadth and width.

Robs hated sweet biscuits, much preferring savoury treats such as the aforementioned chips or curry or pizza. He was, though, a terribly polite dog and never rejected any kind offerings from those that visited, even though he might not eat them. This led to a funny incident with one of the regular Parcelforce men who would collect our parcels for our burgeoning business.

Robbie had become a great favourite of these men who would daily pick up our orders and they all liked their sweets, crisps, biscuits and other snacks that any self-respecting canine would like too. A particularly nice chap once gave him a couple of custard creams and

Robbie accepted the offering most gracefully in his mouth and nodded to thank him, and disappeared into the house.

When I went into the kitchen, I found the gift abandoned in the middle of the floor and, Robs gazing with disgust at the sweet offering, clearly willing me to clear it up! He was quite definite that he, really, really did not like biscuits, under any circumstances. But what impressed me most was that he did not refuse the offering from his friend, preferring to accept the biscuits gratefully.

He was, however, like Ellie, passionate about cheese. I have always been very strict with the animals about approaching the table until the last fork goes down. They were allowed to lie under the table or in the same room as long as they do not move around and there is no begging at all. If there is to be a treat then they are invited to the table.

Unfortunately, one Christmas with my in-laws, Ellie slightly forgot this rule. She has a penchant for getting on a chair at the table, particularly on state occasions, or at parties and sitting very prettily and still whilst pointedly gazing at the cheese and biscuits. It seemed quite endearing and sweet until she did exactly this on Christmas Day a few years ago – she was brilliantly behaved all the way through lunch, lying so still under the table. But as soon as the Christmas pudding left the dining room and therefore there was an empty chair, namely mine, she neatly hopped up onto it and sat very determinedly next to my Father-in-Law, who was desperately trying to look disapproving, whilst holding back chortles of laughter.

She was, however, not indulged at the table, this time, even though my husband sometimes encourages these entertaining antics, much to the amusement (generally) of those present.

All our animals have enjoyed the most splendid culinary feast they can tolerate when their days are drawing to an end, and the rules no longer apply. Some of my earliest memories are of feeding smoked salmon to cats, lamb mince and spaghetti bolognaise, and chocolate and ice cream to dogs, but now the Food Police tell us that this is BAD for them, even if only as a rare treat. It would sometimes seem that the nanny state is determined to remove all pleasure from the infrequent subversive morsel both human and canine.

CHAPTER 15

Barbs, Brambles, Wells And Broken Glass

Down deep in the reclusive tangled wilderness of the woods at Aylesmore, there was an enchanted area where the stream ran swiftly through and the dappled sunlight softly encouraged a wonderful array of woodland vegetation. Snowdrops, wood anemones, and bluebells bloom before the leaves deny the undergrowth of the stronger sunlight. There, Robbie and I used to walk and play at dawn and dusk.

However, as in all fairy tales hidden dangers lurk, and one day Robbie and I received a sharp reminder of this. Walking in an area that we thought we knew well, suddenly, Robs began to disappear from sight down some sort of hole of which we were both previously unaware. The next few moments were confusing, but I reacted so fast that the next thing I knew was that I was holding Robbie by the collar, and hauling back him out of what appeared to be a very deep hole.

On closer inspection, having fetched a torch and Robbie, somewhat insouciant having been rescued, but firmly on a lead, it turned out to be a well some 30 feet deep, presumably sunk to collect water from an underground spring.

Shaken, but in Rob's case not stirred, I was made of less hardy stuff and I vowed to be more careful in future. But Robs was a particularly accident prone dog, and not, I might hasten to add, through lack of care. From the broken paw already mentioned, he proceeded to career through life at the speed of light when outside, which led to all sorts of mishaps.

In Dumfries, in a walled garden, he managed to run through the area where the bottles from the house were discarded, and broken glass seemed to be everywhere. This led to some serious cuts. With blood streaming everywhere we dashed down south to get to our local vet. The journey, stressful anyway, was not enhanced by the highly disapproving stares from fellow travellers at the services. A bloody and cut paw is slow and difficult to heal due to the blood supply and the length of their legs, and this was no exception to the rule.

Even more difficult were the ears, ripped at another concealed barbed wire fence encountered at full speed, and of course on a Bank Holiday, and then becoming infected on the following Bank Holiday. So May was taken up with nursing a very grumpy dog and a large vet's bill.

Dew claws were a speciality. The dew claw is an interesting evolutionary leftover from a distant early ancestor which was a cat-like (don't tell the dog) creature called a miacis which climbed trees. Now often the dew claws of a modern dog are removed for fear of them being ripped as they seem to serve no useful purpose for the modern dog.

Ellie does not have hers, though still seems to retain a remarkable ability to hold things, yoghurt pots and bones in particular, but Robbie's provided no end of trouble, until we decided to rid him of this expensive appendage. I think, also to Rob's relief too.

If there were brambles, Robbie would encounter them at the sprint, if he didn't look where he was turning, he could trip over his own paws. However, he was lucky enough to be protected by modern veterinary science. In earlier days before the welcome advent of x-rays and plaster casts, a concealed rabbit hole could mean a broken leg and the end of a life. This was indeed the fate of a Saluki which belonged to a cousin of my father's. The hound had to be put down when it caught a paw at speed in a hole on Exmoor. The speed which these animals demonstrate sometimes puts their life at peril.

CHAPTER 16

Goodbye To The Enchanted Life

Exile

Dreams have to end somewhere and our Aylesmore idyll ended abruptly and without warning when M was 11 months old. We were given notice: two months to leave. It was as if the world had ended and my husband and I were in shock for a few days as we adjusted to banishment from our precious Eden.

We had had the most glorious times walking on the estate or in the forests around the area, down in the enchanted wood, where Robs and I so often would venture in the early morning to our special place by the stream where the sun beams would spotlight the verdant vegetation. Where the stream would babble away framed by the emerald green mossy rocks and bank, the birds would escalate their dawn chorus until it seemed the whole world was a symphony of light and sound and worship and we would sit together bathed in the promise of the new day.

Aylesmore was truly an enchanted place and we had the happiest of walks exploring the fields, the hedges, the coombes and the views back across the Severn almost directly opposite where I was brought up. Behind our house I had created my first proper garden, week by week, I would take a tiny bit of money £1 or £2 and buy little plants or seeds and watch with wonder as my garden grew: Lilies, Roses, Delphiniums, Lupins, herbs and Helichrysum, Peonies, and tiny Thyme. I haunted the local garden centres picking up the cheap near abandoned sick plants and coaxed them back to life. I loved my garden and our business with parchment flowers which we were importing from Thailand. And then suddenly it was all gone with a single letter from the landlord giving us two months' notice.

Anxiously searching for somewhere to live I roamed the area, finding suitable potential homes, only to lose them through my husband's indecisiveness, and anxiety began to take the upper hand. There were no more of our glorious garden or walks or dreamy perambulations in the early morning through the entrancing wood. Life, it seemed, would never be the same again.

It wasn't. It was different and little did I know Robs and I had many more adventures to come and that some years later I would give my heart again to another magical lurcher.

We moved with an eleven month old baby, a dog and a growing company, to a charming house elsewhere in the Forest called Rose Cottage. Though not Aylesmore it was in its own way idyllic: with a sweet garden, which housed a good number of roses and a large lawn, the cottage stood alone in its own lane surrounded by fields,with the only traffic being the occasional horse or two, or the postman. By this point the warehouse was based at on the industrial estate at Lydney and so I worked from home with plenty of time to explore the area with the dog and sometimes the baby.

Rose Cottage was also blessed with two very charming sitting rooms, one in an L-shape which housed the dining table and another where my piano resided, with doors giving onto the garden. Upstairs there were five bedrooms, two large and comfortable and a long corridor where M at 13 months took her first steps.

Just on the other side of the approach road from Trellech was a wood, the wood I have already described where the sheep cropped pebbled paths lead to an enchanted place where suddenly at the top of the hill there is an open view back across the estuary. This was "The Top of the World Place" which I have already described, where Robbie and Ferdie have their final resting place.

It is an amazing site and one we still visit. I love the walk through a scattered scruffy coppice peppered with small blueberry shrubs and bronzed ferns, along a wide stony forestry clearing, up the majestic beech avenue, where the branches seem to reach the sky, then take a left turn to another path bounded by a crumbling stone wall and lonely wind shaped small straggly trees, until the view is opened up across the sheep cropped fields.

Bright emerald moss and mustard lichen patchwork the old wall and the twin oaks stand leaning towards each other at the centre of the field, looking for all the world like an old married couple. They say that trees talk, these two definitely have the conspiratorial air of a pair long grown used to each other, foibles and all.

Our business was beginning to thrive by this point and my days were often punctuated by the excitement of yet more boxes of samples and new commissions arriving from China. It was there with a young growing child, a very nice nanny and my beloved dog that I designed my first collection of Christmas tree decorations which was to change the direction of the business and begin to establish a recognisable brand. Mornings were spent working from the very early hours but the afternoons were free for walking and exploring, even the time was not wasted for some of my best ideas would occur in that half somnolent state of the leisurely post lunch hours. The evening were fun too and we enjoyed a glorious summer working hard, eating outside and generally enjoying our work/life balance. Robbie too was very happy and cheerful throughout that time and bore the administrations of the over eager toddler with great good grace.

It was a very happy time for us in many ways, a little like a holiday as we knew it was only a short time but I was dreading the impending move to Herefordshire, where we were buying an old house. It had two amazing barns to house the business away from the current premises at Lydney Industrial Estate and great access from the A49 Ross-on-Wye to Hereford trunk road. Life from here would never be the same again.

CHAPTER 17

Gypsy Days

Court Farm was to prove a disastrous move for me as I hated the house and I was working too hard as well as trying to look after a small child. In that time we acquired Ferdie who wasn't the brightest of sparks but a bitch of a very sweet nature, though prone to wandering. Within three years I had left Herefordshire.

Over the next few years, M and Robbie and I wandered like gypsies after leaving Herefordshire, we travelled to south Bristol, then in to Bristol, near school in Henleaze where M was at my old school, Badminton. And then to Olveston, the village next to Aust, where I grew up and finally to Hillesley where Ellie has lived all of her time with us and we enjoy another Badminton, up the road, where the famous horse trials are held.

Seven moves in seven years; it sounds like a fairy tale and in some ways our lives were enchanted, in others scary as if we were in a dark wood. My father died, Robbie died, I had next to no money, I lost my bearings and finally in 2007 my final lurcher, Ferdie also died.

At least that is how it seemed at the time. FINALITY. The end of all things. One thing I can tell you, though, is that we seemed to be protected by fate at some level.

In the intervening years I owned a shop, and M was growing up. She attended three different schools in the same period and a further three before university. My father's death came as a terrible shock; death is always shocking however accurately foretold but he died of cancer 21 days after diagnosis. My brother had lost his father-in-law some months earlier, and my now husband, his mother totally unexpectedly and at a very young age the previous summer – so it seemed as if the order of the world was turned upon its head and weddings and christenings were replaced by funerals and memorial services. Truly we were all growing up and this time was a rite of passage. Unbeknownst to me around this time I met Christopher, my husband at one of these funerals which led some years later to Ellie coming home.

CHAPTER 18

Pneumonia, Death, Near Death Experiences And More Doggerel

In the intervening years when I had no dog I contracted Pneumonia; I had been out in Italy with friends for Christmas 2007 and on December 23rd I had been walking in Siena, in the cold wet rain, around the famous Palomino with holes in my shoes, which was very stupid as I had already flown out there with a serious cold. I had just bought a beautiful coloured print for my daughter when I realised that I was seriously ill.

To cut a long story short, that was the day before Christmas Eve, on Christmas Day I was violently sick, on Boxing Day I flew back to England. Even the taxi driver said he should be taking me to hospital, I said no. I could cope.

I couldn't – I arrived back to find Misty, the cat, had died, even though she had been taken great care of by a kind neighbour, which was very traumatic for him. M was very upset on her arrival home. Britain was in the grip of an icy cold spell and I was not well enough to take poor Misty's body to the vet so I am afraid she was put in the shed where she froze solid.

On New Year's Eve, I conceded defeat, I could not cope and realising that the doctor's would be closed for the next few days, I drove myself to the surgery very early and presented myself at reception. Greeted by the angry glare, which the reception dragons seem to wear when confronted with any potential patient who might bother the doctor, I unintentionally hastened the process of an unscheduled appointment by proceeding to collapse.

"I need a doctor," I gasped, fighting for air.

Suddenly reception sprang into action.

"Someone get a doctor!"

Well, I saw a doctor, went for an x-ray at a near-by cottage hospital, was confirmed to have pneumonia and given co-codamol and doxycycline and went home to rest. My mother came to look after M and I do not remember that New Year at all.

Unfortunately, we did not realise that given my coeliac disease I should not drink whisky but it had always been a staple cure-all for all colds, flus and other bronchial issues at home, so nobody thought twice about the odd wee dram to ease the pain.

Big mistake huge – the combination of an opiate (co-codamol), a hefty anti-biotic and the whisky, which not only was alcoholic, but made from malt so I reacted to the grain (we did not appreciate how extensive my coeliac had become) was near lethal.

The days dragged on through January, my mother taking M to the school bus and generally she was around, though out of the house a lot. I might occasionally surface in the evenings for a few minutes but I did not want to eat anything except spinach and would drift apathetically back to the confines of my room.

I am an avid reader and spend huge amounts of time on my computer, generally, but I do not remember doing either of those things. I recall sleeping a lot and a great deal of pain and times when I couldn't get my breath and I would be all alone...

I shall never know whether the following experiences were real, imagined, induced by drugs, or even self-induced but one was very frightening and the other incredibly reassuring and, in some ways, wonderful and amazing.

The first was nasty, I was fighting for air and it seemed to me I was drifting in and out of consciousness, as if I was in a grey fog and I could not quite get my mental bearings; the most curious thing was my mental distance from myself. Not only was I highly aware of my physical state, but I found myself observing my own mental state, as well, in an interested state of detachment.

The following experience may well have been the product of an over active imagination and or a disordered mind due to illness, drugs and alcohol but I saw, with the accompaniment of Chopin's Raindrop Prelude, which I could hear quite distinctly in my mind's ear, a procession of monks travelling across my bed. Hooded, black-grey and sinister, the queue filed lugubriously and slowly as if in a swirling mist and all colours of grey and off white; the crowning terror was the final spectre in the sequence, who turned to look at me, he carried the scythe and his face was Death. Even now, more than ten years later, I can recall with extreme clarity, the terror that clutched my heart and mind.

I do not remember any further other than catching my breath.... and then nothing.

I tell this grey tale simply to point up another experience that I had during that time of illness which is actually more fascinating to me.

Many years earlier I had been talking, as one does to dogs, to Robbie who was beginning to show his age and I was very aware of death at the time as my father had recently died of cancer, in 2003. I said to Robbie, aware that he was likely to die before me as he was approaching 12 years 6 months at the time,

"When I die, Darling, I hope that you will greet me at the gates of heaven."

Robs turned his wise old and loving eyes towards me and, perhaps fancifully, I detected a nod of understanding and comprehension in them. I did not refer to the incident again as I hated to dwell upon the time when my darling boy would no longer be my shadow glued to my left hand side and forever present. When he did indeed die in 2005 it was probably the most intensely sad time of my whole life and, although I had the lovely Ferdie, she, too, was grieving in her own way, as was M. So it was a dark time for us all.

And so here I was in January 2008, Ferdie was no longer with us either and I was struggling for breath in a darkish room. It must have been approaching sundown on a grim day, the house was empty and I felt distinctly odd.

Suddenly I saw blue sky and a rush-seated ladder back chair with my father seated upon it, almost hovering in the sky with fluffy white clouds scudding behind him, and he was smiling.

Then most unexpectedly, from the right hand side of the chair as I was looking at Daddy, Robbie snaked around the leg and looked straight at me. his eyes full of compassion and love.

My father lent forward and stroked Robbie, which was odd, since they had not always got on during their lifetimes, and spoke very clearly, firmly and kindly.

"You still have work to do."

This sounded frighteningly clichéd to me, for even in my presumably delirious state, I was aware of the possibility that all this was the product of my fevered imagination. I was aware of the fact I had asked Robs to meet me when I died, the fact I did not want to die just yet because I had not yet made a difference to the world, my father was perched uncomfortably on a strange chair in the middle of the sky, and I knew I had to get better.

The vision faded and I remember feeling intensely almost insanely sad, worried also because it was a possible indicator that I was more ill than I had considered. However, I was filled at the same time with an awareness that there would be a life beyond the greyness and solitary isolation of severe illness. And that one day I would be able to walk again outside in the fresh air and live a normal healthy life, even though that seemed a very long way off at the time.

It seemed that my father and my dog had come to me in my hour of need and gently reassured me. Though it took another five weeks before I was able to leave my bed, somewhat under duress, for my mother left in late February to go and live with my brother in Somerset. Consequently, I had to get back to doing the school run and other parental and domestic duties.

2008 had started really badly. It was a tumultuous year, but I met my husband again in May of that year when his persistence paid off and eventually our engagement and subsequent marriage led to this story: the story of Google Girl. So Rob's intervention was instrumental in returning me to a world where I could have the joy of a wonderful dog and life again.

It has not been plain sailing at all since then, obstacles to our marriage, family feuds, Ellie arriving in 2009, various severe illnesses on my part due probably to coeliac and other problems, generated by things largely beyond my control, have plagued these years. But now writing this, I can see how much we have all weathered and survived.

I hope that my father and my dog will be proud of me now as I piece together again the patchwork of our lives into a new and brighter cloth and a glorious future. And Ellie has been so much a part of learning to live happily again, to have had one really special and extraordinary lurcher is lucky, to enjoy the companionship and love of a second legendary animal is beyond my wildest dreams.

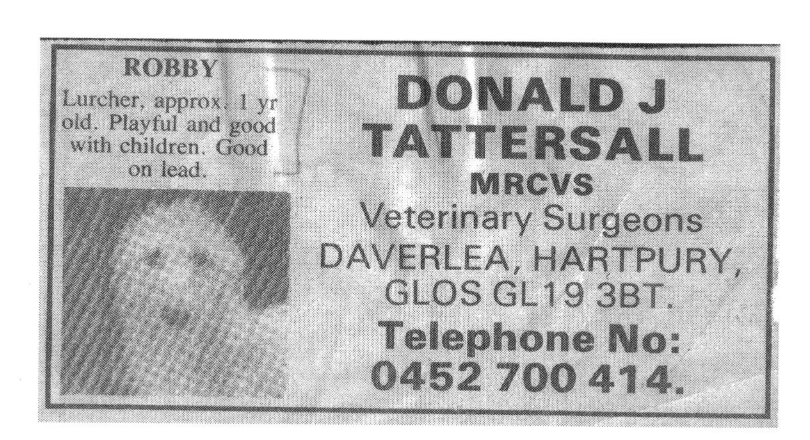
ROBBY

Lurcher, approx. 1 yr old. Playful and good with children. Good on lead.

DONALD J TATTERSALL

MRCVS

Veterinary Surgeons

DAVERLEA, HARTPURY, GLOS GL19 3BT.

Telephone No: 0452 700 414.

The Advertisement

Robbie Soon After He Came To Us

Getting Stronger

Happier and More Confident

Alert and Listening

Cool, Calm and Collected!

Poised

Cool Dude

Snow Dog

With Amor at The Old Parsonage

Looking East Across The Severn Estuary

Snowboarding

Two Dogs

One Dog

No Dogs

The Adieu

The Old Parsonage and St. Augustine's Vineyard In 1993

Aust Beach - Looking Across The River Severn To Beachley

Guinea Fowl

Ermentrude and Chicks

The Earl of Essex

Fly and Robbie at The Old Parsonage

The Lurcher and The Grapes
What's The Moral of The Tail?!

The Stream at Aylesmore

Aylesmore Snow - Land of Enchantment

AMPHIBIANS, CATS, BATS, RATS AND FEATHERED FRIENDS.

CHAPTER 19

CATS

Ciao was the first animal I ever met. A cat of stunning character and resolve, he was my friend and self-appointed protector. He was also prone to mischief, and fearsome attacks on those who displeased him, as well as being an accomplished thief.

He was my Mother's fearsome Siamese: beautiful, stealthy and cunning, bringing to mind TS Eliot's "MacCavity", for Ciao was quite clearly the "hidden paw" from "Old Possum's Book of Practical Cats". Oddly enough he was probably also the flying squad's despair for my father was an engineer who had recently left Westland Helicopters at Yeovil when he met my mother.

On their first date, Daddy came to pick my Mother up and received a very unwelcome welcome from Ciao, who unleased himself upon my father's nose with quite unexpected and undeserved ferocity.

After my father was stitched up, they still went out to dinner and in due course became engaged. They married on 3rd September 1963, and often joked that it was the anniversary of the day war broke out. I was born the following August.

Being the 1960s, it was the custom to leave babies outside in the fresh air and as far away from the house as possible, so you couldn't hear the baby's wailing. Apparently I was a difficult baby prone to long outbursts of crying. Ciao was often to be found lying with me in the pram outside and was clearly very protective of me in my early months. Even so, my Mother was absolutely mortified to go to check on me in my pram outside and find Ciao perched in the pram feeding me mouldy stilton he had clearly got from the rubbish. Not for the first or last time, he would scavenge for me, clearly having worked out the process for himself and concerned for my welfare.

Ciao was a thief too. My parents recalled with some horror not a little amusement that once they went to a local dinner party only to hear an embarrassing story about vanishing food. The hostess described how she had lovingly prepared some fillet steaks for her husband and herself for dinner as a special anniversary treat. She had left the kitchen momentarily to spruce herself up for supper, leaving the beef to rest and the redolent smell of a well prepared dinner. She came downstairs, all glamorous and eagerly anticipating her husband's imminent arrival only to discover to her horror that everything was in its place but for one thing: the steaks.

She recounted this story with great glee at the party, as luckily her husband had already booked a table for surprise dinner elsewhere so they had a lovely evening nevertheless.

My parents exchanged looks as discreetly as possible: the day of the theft was two days previously. My mother had forgotten to feed the cat and he had shot her one of those looks which says "***** I'll go and get my own supper". Some minutes later Ciao had arrived back with a couple of steaming steaks in his mouth and looking very pleased with himself. He swore at my mother when she tried to remove his bounty and enjoyed a dinner fit for a king.

My parents left the party as soon as was possibly polite, but never in a million years had they expected to hear how the mysterious steaks had appeared.

Ciao was also determined to be a cat of a different variety; an independent character, as you may have already determined, he was prone to his own idiosyncrasies.

One day my mother had made up a large batch of Seville orange marmalade and laid out the jars to cool in the kitchen and went elsewhere to do something else. Sometime later she came back to find every jar had been dipped into. I was still too young to reach the counter so she could only come to one conclusion: the thief was at work again.

Following a sticky marmalade trail which led outside to the pond she found a Marmalade Cat sunning himself in the summer sun and licking his paws furiously, not sure of the taste, but a very sticky orange mess. He was peremptorily marched off to the bathroom and cleaned up post haste. Needless to say the marmalade was bottled and still happily consumed by all, unknowing that, the CAT had been there first.

CHAPTER 20

The Burmese Twins And Sadness

The successors to Ciao, whose death I do not remember, were a pair of Burmese twins of the highest pedigree, who my parents called "Worcester and Soy".

They were the most enchanting pair you can imagine and I am quite sure that my parents loved them more than we children as they were slightly less difficult. Every night I would mount the stairs with a kitten under each arm and cuddle up to them as I fell asleep. Usually they would still be there be in the early morning when I awoke to the sound of their contented purring all curled up at the bottom of my eiderdown.

They were the deep brown Burmese with the incredible yellow green eyes. Burmese cats were renowned for being trained by the Burmese long ago as warriors to be carried on their shoulders and to attack the enemies' eyes. These two had clearly forgotten their fearsome warrior roots as they were as gentle as anything toward us humans and in particular children, but they had retained the atavistic hunting instinct.

This was to prove a great sadness for us, for as they grew older, rather than sleep upon my bed, it became their mission to hunt at night. Having catflaps of course they were free to come and go at will and one night disaster struck.

We lived at the side of a busy road in Bristol at the time and their favourite game became "chicken" or "cat chase across the road. One fateful night, early in the morning, Soy was the victim of a fatal accident, entirely not the driver's fault.

It is the first time I remember my Father crying. Both of my parents cried for days and it is the first real experience of loss and sorrow that I can remember. I would have been around three or four years old and had lost a friend and a companion too. A few years later we moved to the country, with Worcester and soon began to acquire the menagerie I shall tell you more about. It felt like a legacy to Soy who could hunt forever in Heaven. Worcester as you will read went on to be a real hero and here the real adventures began.

CHAPTER 21

The Blackbirds

When I was a very little girl, whilst still at Falcondale Road, I told the most shocking lie. Except it became true.

"Mummy, the blackbirds are coming to feed at my window".

Of course they were not, but I wanted them to. The very next day they DID come to my window and looked me in the eye. They came every morning and evening and sang me to sleep. I would pinch food for them from the kitchen and leave it on the windowsill. It was probably around this time that I started to closely study birds and animals, an occupation which has become a habit and an enduring passion.

The blackbirds continue to be part of my life as every year we have at least one breeding pair in the ivy in the garden. Sadly often the young parents lose at least one brood to neglect or lack of knowledge until they finally manage to raise a family. This year (2018) we lost a most delightful and tame little chap who we had witnessed from an egg. I fear he may have fallen prey to slug pellets. We were very saddened by the loss of this special young life.

The sight of the young birds on the ground fat and fluffy, more like baby cuckoos than their elegant and frequently tetchy and exhausted parents, demanding more and more food reminds me of spoilt children in the supermarket. It is amazing how these cartoon like babies grow into the sleek and beautiful adults with their elegant and pure songs which change, mark and denote the seasons.

The only time of year we tend not to hear the blackbirds is in August and for years I wondered why until one day I worked out that they must go on holiday to replenish for the winter. Here they go to the woods where there is an abundance of ripening berries and nearby grain fields, and so the once exhausted, harassed and skinny parents restore themselves. They return to the garden in September when their autumn song marks the turning of the season and harks to the approaching winter. Even now as I write early in the morning I have just heard the first uplifting notes telling me that the rosy fingered dawn is visible and morning has broken…

It can however be irritating as the youngest every year wake me with their inimitable calls and all the adults tell them to shut up…. Every year is a different brood and every spring I am woken up at the first juvenile squawk and the parents telling the babies to go back to sleep. This year there was a particularly aggravating youngster who took it upon himself to squawk as loudly as possible at least 45 minutes before anyone else. Sometimes I find it just as irritating as the adult blackbirds and long to return to slumber but the music is insistent and then suddenly the whole dawn chorus stirs into life and there is no going back to sleep. It is time to prepare for a new day, for morning had broken.

CHAPTER 22

RATS

In 1974 we moved to an old house close to the River Severn. It was around 200 yards from the actual river but on a tiny little hillock which seemed to belong to the house itself. Although the former parsonage for the local church, it had been renamed Coleshill until my mother took action to rename it "The Old Parsonage". It was in the direct onslaught of the oncoming winds and weather from the Bristol Channel. Virtually derelict when we moved in, there was no running water nor electricity, and we had to draw water from the well which was at least 30ft deep and still remains to this day. In later years it was a great temptation for the grandchildren to deposit any keys left lying around the house, much to my father's anger.

There were a number of unwelcome incumbents: fleas, flies, mosquitoes, and most of all, RATS. (I think the mice had hidden.)

The house was a noisy house and groaned and creaked at night, due to the prevailing south-westerlies. I hear in my mind to this day the sound of the fog-horn warning the ships in the channel up to Sharpness, "Be careful, be careful", a sound that would punctuate my sleep.

It was idyllic in many ways, we had space, we had freedom, we had views, but we inherited some very unwelcome tenants.

One morning I went downstairs to find my school jersey which I had left in the study and picked up what I thought in the darkness was the sweater.

It wasn't – it was a DEAD RAT!!!!!

Our Burmese cat, Worcester, to whom you have already been introduced, took it upon himself to rid the house and grounds of this verminous infestation and so every day, two or three times a day, he would proudly bring in his latest conquest: unfortunately I had picked up the very latest….

But worse was to come….

Early one morning at around 4am I was awakened by the sound of my parents' talking: my father was due to leave for Germany by an early flight from Heathrow and they were waiting for his driver. Alerted from sleep by the sound of voices, I went to the bathroom and by chance looked out upon our stable yard, or more specifically the coach house roof which was wreathed in the early morning mist.

There was the most unusual and horrifying sight I had had ever seen.

On the roof of the coach house there was a gathering of the largest number of rats I have ever observed: in concentric layers they ranged from the largest in the innermost ring and through four or five circles to what were clearly baby rats on the outer edge. The adults seemed to my childish mind to be deliberating and the sight was just too scary and strange…I screamed for my parents who came rushing up the stairs into the bathroom just in time to witness the most extraordinary sight.

As the sun just began to hint at rising, the circles parted and from the nucleus of the gathering came a huge rat, a King Rat, followed by the other males. He led off across the roof, down the drainpipe, over the compost heap and over the stone wall into the road.

He was followed by all the other rats we had seen on the roof, in strict formation, one by one they left the property in a long queue like refugees. Worcester had won! I only ever saw another rat again on the land years later but it swiftly retreated.

We had caught a glimpse of a rat's parliament, spoken of in legend, written in song, deliberating about moving to another home. They had been defeated by one single brave and persistent cat. The local farmers didn't like us much as they inherited the rats.

Years later in Herefordshire we had a major rat problem in our outhouses, but one day I spent a little time watching, with enchantment, young rats, who had been born in our woodshed, playing gleefully in the newly fallen snow. From then on we decided there was no poison, and oddly enough there was no more problem from the rats. Many years later, I cleared a property that I was renting of another rat invasion by the liberal use of Peppermint essential oil and prayerful intention. Amazingly the rats retreated once more without a fight or a single death.

What one cat did was exceptional. Worcester cleared the whole area systematically and with precision. What one rat did was the same. He saved his tribe and our cat sought to protect us - such is the balance of nature and fealty.

Actually, there is an addendum to this story: here in Hillesley, we had a charming rat who lived below the shed. He was very sweet and loved engaging in the washing of his paws just outside the house where he would bathe in the sunshine and smile at me. One day he left and I never saw him again. I had nicknamed him Templeton after the rat in *Charlotte's Web*. So don't misjudge another animal just because you are frightened, just understand we are all different and never ever underestimate the intelligence of the rat.

CHAPTER 23

SWALLOWS

Swallows have flown through all of my life – they played with my dogs, irritated the cats no end and I have watched generation after generation grow up.

One particular family used to come and knock on the kitchen window whenever Worcester, that great Burmese Cat hero, would go into the outhouse to taunt the fledglings in the nest and try to get up to that nest, which he never managed. Naturally, I would run out into the courtyard and shoo the cat away. He still slept with me though, often climbing the wisteria to my parents' room, howling to be let in and then leaving them, that is until the dog arrived…

One year in my twenties I was severely ill after leaving my job at an American bank, so had to return home to recuperate over the summer. I spent many hours observing the young swallows on the house telephone wire which we could see from the window seat on the landing.

It was fascinating watching the little ones being fed, both parents busily, and it seemed endlessly, flying to them with flies, back and forth, ducking and diving and flickering and fluttering. It soon began to be apparent that one of them was growing much faster than its other four siblings and the evidence was in its very different size. I began to wonder why and started to scrutinise the family even more closely.

The bully boy was clever, the parents were feeding the young sequentially but he would move along the line so that he was ALWAYS the first in line. It took some weeks for the parents to notice, to my growing concern, then one day, suddenly his father swooped in like bomber command and knocked him decisively out of the way.

The chubby little fledging appeared astonished, and not a little shocked. After that discipline was kept and they all began to grow equally with the others catching up. That year I cried when the starlings came reeling like tea leaves against the sky, for that means the end of the summer and that the Swallows have to leave. We had a golden cocker spaniel called Saffron who loved to play with the birds in the field and they would dive bomb and tease her, with both parties clearly enjoying the fun. But one day she went out into the field and barked, and barked, and barked, but her swallow friends did not come and she was terribly sad for the next few days. The next summer, though, she was a bit savvier and resigned to their unannounced departure.

Adieu, my friends, until next summer!

Here too, at Hillesley, we have swallows, house martins and swifts, and one September I saw the most extraordinary sight of a large group of house martins feeding one single nest; a sight I have never witnessed before in over 40 years of watching these bewitching summer birds.

Early one morning around 6:30 am I was chatting with a neighbour about to take Ellie out for a walk when we noticed a lot of chattering burbling noise and up to twenty house martins swooping up to a nest just under the eaves of a house just across the road.

Fluttering and swooping like little missiles, this unusual group were consistently hitting their target of one nest; normally one would see the immediate family feeding a nest, about three adults, the parents and perhaps everybody's uncle, the one who didn't get a wife! But here was the extended family, aunts, uncles, cousins maybe, concentrating on one group of maybe three or four fledglings (we could not see how many were actually in the nest).

I can only surmise that this was a second and late brood, and with the autumn approaching fast and the babies not yet fledged, there was some urgency about the situation as they were already preparing to leave in two or three weeks and this little family had not yet been through flying school.

After that day, interestingly, I did not observe any activity at the point, so it possible they fledged just in time to learn to fly south. Let's hope so.

Flying school is a magnificent sight to watch. Having observed it often as a child, I can now enjoy from the window seat in my study the wonderful achievement of these young birds, particularly the house martins. Both swallows and house martins share remarkably similar teaching rules and quite different from other bird groups.

Having coaxed, scolded or bribed your babies from the nest, apparently you line them up on the nearest available roof top, and then show them, by example, how you deploy your wings and trust to the universe that you can fly.

Some of the braver little birds take to the skies with great aplomb and are soaring away and showing off within hours of first take-off but it is the more timid ones that are sometimes more interesting to watch.

This year I watched one group of four; three were happily playing tag in the sky after only a few hours but one was highly reluctant to trust mother, wings or nature and stubbornly remained on the ridge of the roof refusing to budge, despite repeated efforts by the mother to dislodge it so it HAD to fly.

This stand-off went on for some days and I watched with some amusement but more with concern, the mother and father become more and more frustrated with the youngster who was clearly very stubborn and firmly of the opinion he was not born to fly…

After some agonising hours one morning, and some very cross and anxious parents, the little one managed to fly back to the nest without coaxing or bullying and suddenly it was like a light bulb came on in his head…Later that day all four siblings were airborne, much to the relief of the adult birds, I am sure.

It is fascinating to watch the social development of these clans. Over the years I have noticed that whilst they start off in their family groups, they then coalesce with the other youngsters in flying school and are supervised during the day by fewer and fewer adult birds, who seem to go down into our nearby valley where the flies are more plentiful and there is more water available in open streams.

In the evening we see "cocktail hour" when the adults return, and in the time before bed there are up to a hundred martins dancing against the colourful backdrop of the setting sun, weaving and ducking and diving against the deep orange, red and purple of the sky with the outline of the Welsh mountains in the far distance. It is a sight we feel so privileged to enjoy.

As the swallows and martins gather on the telegraph wires and the chattering becomes louder and more urgent, we know that they will leave soon and then for a few days the skies will seem very empty of their glorious ballet. Just before they leave the starlings start to gather in swarms around the church roof, it is as if the aristocrats are to leave for their winter residence, and so the more humble starling is allowed the freedom of the sky once more. The swallows and martins generally leave here around the third week of September, though we do see some stragglers from up north and Scotland into early October and then we await them from the middle of March, although they have been as late as April. This year, 2018, they came in early May, delayed perhaps by the monumental snow in March.

The enigmatic swifts are already long gone by this point of the year, their noisy presence is really only noticeable in May and June. Until this year I had never seen a very young swift as they tend to do their screeching bombing raids around dawn and dusk so the light is more limited to observe them. However, this year I watched in awe some twenty-five juvenile swifts flying in formation and uttering that inimitable cry. I have never seen so many at once and during the day too, so I am hoping that this bodes well for our local Swift population.

When I lived in Herefordshire, our evenings were enlivened by large groups of swifts who would come up the valley at very high speed as if targeting the back of the house. Suddenly they would adjust their flight angle sharply upwards at the very last minute up and over the roof top, all the time screeching in that idiosyncratic manner. It was like witnessing a Red Arrows display or being at an air show each evening! It is worth noticing their top speed approaches 70 miles per hour; in 2012 a speed of 69.3 miles per hour was clocked, faster than the average speed of our motorways and probably with less disruption.

They say that swifts do everything on the wing and never rest.... It is also said that a grounded swallow cannot re-launch which I know quite categorically that is not true as I have seen them launch from the ground. I have also returned a fledgling to the nest in our outhouse when I was young and the parents accepted it back even though it had already learnt to fly, and I had touched it.

What wonderful fantastic flying machines this group of birds is and what a joy and excitement they bring to summer and long evenings of entertainment to say nothing of the stubborn youngsters who are quite determined that they were never meant to fly and suddenly discover the majesty of their inheritance, the skies.

CHAPTER 24

Of Newts, Frogs And Other Amphibians

At the Old Parsonage, we had a cellar which had two very steep sets of steps: one out to the garden which was built of red brick and totally treacherous, and the other within the house which was made of stone and wound down to the underpart of the house through which a stream ran.

This later became known as the hobbit hole, as my father had his desk down there in the wine cellar when we had a vineyard, but in the meantime we had some interesting inhabitants. A newt and his family resided in the stream which ran through the cellar; they lived in the drain and occasionally could be seen running across the floor. Another resident was a somewhat stately Toad who developed a most peculiar habit. When he heard the inimitable sound theme of "Coronation Street" on the television (don't ask!), he would hop up the steps (about fifteen of them) and stay for the entirety of the programme until the closing credits. Then he would descend back down to his home in the dank dark cellar. We never knew why and I am sure that the cast of the programme would have been astonished to know they had an amphibian fan.

Here in Gloucestershire, we have a tendency to leave the Garden Room doors open and early one morning I was tidying the woollen throw from the sofa when suddenly a frog leapt out of the blanket. I don't know who was more surprised, the frog or I.

Clearly it had come from the pond but we have never been sure whether it came into the house of its own accord and curiosity or whether it had been uplifted by Ellie and carried inside by her. Whatever the provenance, the frog made a rapid and somewhat undignified exit from the house, then leaping into the pond with all the agility and sheer athletic ability of the mightiest Olympian...but not before giving me the most fearful shock!

Frogs seem to have played a curious part in my life even as a very young child. The garden at Falcondale Road was magical, despite being attached to a house at the side of one of the busiest trunk roads into Bristol. The front garden was Mediterranean in style, sloping down to road with paths cut through the rock and almost southern European vegetation, which served very well as a jungle and a wonderful exploration and imaginary expedition space, but it was at the back of the house where the real fascination grew.

Quiet and secluded from the traffic, there was a rose arbour abundant in summer with scented blooms, under planted with pungent herbs, and surrounded by tall trees. It was difficult to believe that the road was so close. Central to the garden's appeal was the pond which teamed with life and became an endless source of fascination for my young mind.

There I mainly remember the frogs, seemingly hundreds of them every spring and summer gracing the seasons with their songs and voluptuous reproduction. Every year we witnessed the miracle of life reborn from jelly to full grown amphibian. Every year the frogs would apparently disappear for a while probably to the vegetation around the garden and then return annually celebrating the reunion with their inimitable crooning filling the air in spring, their songs of courtship as distinctive as any southern cricket.

Frogs are highly responsive to human attention: if you sit by a pond quietly at certain times and sing softly the frogs will emerge from their hiding places and sit and listen. Sometimes if you put out a hand they will sit peacefully in your palm until stirred by some atavistic fear, they suddenly leap back into the water.

They are right to be afraid, inadvertently many times as a small child, I unintentionally hurt or even killed the frogs through my fascination with them, by picking them up to examine them when they were too young and fragile to be handled. Being young and naïve myself, it took some time before I recognised the harm I had been doing. The only forgiveness I can give is that I was genuinely too young at first to know any different and when I did, I stopped the unintended torture. The guilt haunted me for years to the extent that at thirteen years old at school, I flatly refused to dissect a frog, one of the staples of the British Biology curriculum, and stormed out of class. (Thereby discovering that it was possible to skip lessons without being disciplined, a discovery of which I am afraid of I took full advantage.)

Luckily for me, this unfortunate interlude, though casting a guilty pall over my early memories of amphibian society, did not affect my good relations with the frogs in general, either in that pond or other watery oases later. I have been so fortunate to have been able many times in my life to enjoy observing these wonderful creatures at close quarters and so you will discover has Ellie.

The pond here was a true test of my husband's love, as when I mentioned, without any particular intent, that it would be "nice" to have a pond as a central feature, little did I know that I had unintentionally handed Christopher a mission.

The next weekend he arrived armed with information about ponds and liners and pumps and I listened with growing astonishment to his plans. I drew an outline plan of where the pond should be and the proportions and we started to measure and then he started to dig, and dig, and dig, and dig...My father-in-law joked that Christopher was digging my grave and indeed it is just the right size, but actually, it is a small area that teams with life - we have frogs, pond skaters, toads, and even one year a pair of tiny newts who we saw through clear water in the Spring.

Cuddling together at the bottom of the pond they were barely an inch long but the larger of the two, presumably the male, was not afraid of the giants peering through the water from above. He merely waved us away with great dignity, looking at us straight in the eye, and somewhat chastised we retreated from our voyeurism, if indeed it was.

We also have dragonflies in August, butterflies in the buddleia and even one year a whole colony of bumble bees in the cavity of the outside wall of the garden. We could not work out what the buzzing sound by the Sky and then we noticed how many bees were coming into the house and went to investigate.

We discovered a small hole in the wall just behind the wall by the television, into which my husband stuffed some paper, in an attempt to deter the bee invasion. Big mistake, we only then acquired some very angry bees indeed as they swarmed around trying to enter their home. I took pity on them and perhaps foolishly removed this stuffing, bare armed, whilst praying that I would not be stung by the large number of bees dancing in anger around me. Protected by intention, perhaps, I emerged unscathed and we had détente.

This year we have no bees living there…but masses in the garden which is a vast improvement on previous years as the population had been in noticeable decline.

Truly this little garden is an eco-system in itself and it is marvellous, especially in the early morning, to witness the spectacular profusion of life in such a small space. Sometimes the sparrow ghetto in the rose hips and the birch are so noisy, that I tell them to shut up which works for a millisecond, until one of the cheeky little urchins hops onto the threshold of the garden room and the twittering raucous chirping starts afresh. Many is the time we moan about our sanctuary, especially at summer dawn, but truly we would not be without our noisy neighbours.

South-east facing, we have the benefit of the dawn views and sunlight in the summer, until five or so in the evening. We are lucky that when the sun comes off the back garden, it remains warm due to being a walled garden. We can witness the spectacular sunsets from the window seat in the study which faces north-west over the Severn estuary, with far reaching views, on a good clear day, to the Brecon Beacons and the Black Mountains.

Many is the time we have watched the sun go down and spread its colourful brush around the whole sky which pulsates with vivid pinks, yellows, reds, oranges and that deep violet peculiar to the sky. The sun appears to sink slowly down and then suddenly, just before the final glimpse, it is as if the whole world is suddenly hushed and awed in the penultimate second before true dusk is born of the twilight.

Then the sun is gone, and soon night falls. and then the sounds change in the garden to love songs on warm nights, especially in the pond which is lit by solar lights, the legacy of the sun during the day. On balmy nights, the air trembles with love and joy and enticement, and Ellie is fascinated by this alien world and spends much time frog hunting and staring at the inky darkness of the teaming pond.

Ellie is otally intrigued by the frogs, particularly in the spring, when they use the white irises as their marriage bower and the nights are filled with the sonorous calling. Some of them are very tame and will sit on my hand. And many is the time I have had to reassure some frightened creature that has been intimidated by Ellie's interest that everything is alright stroking its back in the palm of my hand until it leaps back into the water and harmony is once again restored to the pond.

CHAPTER 25

Seabirds On The Shore

"First there were two of us, then there three of us,
Then there was one bird more,
Four of us wild white sea-birds
Treading the ocean floor...
And the wind rose, and the sea rose,
To the angry billows roar-
With one of us – two of us – three of us – four of us
Seabirds on the shore."

"The Storm" by Walter de la Mare (1893-1956)

We had had a great storm the night before, it was early October when the storms often sweeep in from the Bristol Channel, and the following morning I went down to the warth by the river Severn, with Benjie, our first lurcher. Walking along the beach, sensing the white spume and grey sea spray, I watched Benjie investigating, in that peculiar manner of the canine, the base of the famous and dangerous cliffs seamed with different geological periods of burnet clay, granite rock and white crystalline quartz sandwiches. Aust Beach is not really a beach, and it is a very dangerous place, especially near the cliffs where daily evidence of the rock falls and the coastal erosion is visible along the base. Although there is some underlying sand, it is pitted with rocks and crystals and stones of all varieties, a veritable testament to the geological significance of the area.

Aust Cliff is a 5.30 hectare site of special geological interest which is the most productive Triassic site in the UK. Here a plethora of dental, reptilian and piscatorial remains can be found. Although heavily visited and pillaged by visiting fossil collectors there is always something new upon the beach not only because of the instability of the highly dangerous cliff where the structure crumbles or falls at any time without warning, but also there is also the flotsam and jetsam of the tide. Sadly, this now is often a testament to our consumerist times and the horrific legacy of plastic which plagues our oceans and tidal areas. The dogs, however, have never cared anything for this, what they love to do is potter through the mud marsh and run.

Benjie was snuffling amongst the falling trickle of stones in the background and the nearby fresh water streaming toward the sea feathering tiny tributaries in the sand, when suddenly he took my hand very softly in his mouth. This was a favourite form of communication of his and meant "Look at this" or "Pay attention".

Gently, and near silently, he guided me over to some of the fallen rocks so close to the cliff they were technically in the no-go and unsafe zone, and pointed his nose just underneath three tumbled stones. He raised his head to look me straight in the eye and then tossed his nose again in the same direction. I bent over to investigate: a baby seagull was nestling under the rocks, quietly and almost motionless. Clearly the youngster had been caught up in the furore of the wind and river.

The beach was lined with debris, the flotsam and jetsam of the tempest, and we could not abandon this young creature, who by instinct or parental design was sheltering in the aftermath.

The baby seagull had a broken wing.

"Benjie, Please if I take the seagull for help will you stay close on the road, because I cannot hold your lead and take the baby bird at the same time?". Benj responded with a nod of his head and he was never more obedient than on that walk home. I stuffed the baby gull inside my Barbour jacket and set off home, a little unsure of the welcome my little injured acquaintance was going to get.

When I arrived home I found my parents upstairs painting a bathroom, I told them that I had an injured seagull downstairs. My father's response was less than enthusiastic and he swore that he was not paying for a flipping seagull to go the vet to which I immediately retorted "He's not flipping or flapping or flying or anything at the moment."

My mother however, downed paint brush, immediately rallied round and we rushed the poor little thing to the vet. (These dashes to the vet were always somewhat alarming as my mother was prone to picking up a police escort along the way due to her very positive and robust and driving style on such occasions!) The vet examined him and took him in immediately to fix his wing and over the ensuing days he ate them out of sardines and house and home. I was waiting with increasing trepidation, daily anticipating the mounting bill.

Gulls are extraordinary birds. They seem to be highly intelligent scavengers and I am often struck by their scouting abilities when you see the one or two seemingly solitary fliers hovering inland, only to spy later, huge flocks of them swooping, and fluttering, and squabbling, and bouncing, and marching behind the ploughing tractor, and pinching the seed.

They sometimes appear to be the hit and run burglars of the avian world and are well known for breaking an entry into the seaside shops and exhibiting a huge talent for airborne shop lifting and bin rustling. Often they are to be found over at the local car park, when they sort through the rubbish bins for the most delicious leftovers, or picking up the stray abandoned chips. Their worst trick is sneakily pinching the ice cream of some poor traumatised little mite whose face is instantaneously festooned with the most unbecoming tears and wailing until placated by some tetchy adult.

The little gull's recovery had been excellent, he had become the darling of the practice, if not their greediest patient, and in due course he was released back into the wild. Thrilled by his recovery, I was also immensely relieved that to be informed by the wonderful veterinary surgery there was no bill to pay. Apparently bills are for the benefit of ownership and they treat wild creatures for free (I wish I had known that before!)

On these shores I experienced the reeling, weaving sight and cacophony of the many different birds feeding upon the estuarial shore. How anyone can describe the squabbling, hooting, honking, calling, flapping wings, quiet cooing, busy beaks, and the sound of squelching mud or the rush of impact as the birds hit the water or the beating sound of the flocks taking off, as silence is beyond me. Not even peaceful, but so rich in texture and the living legacy of our coast.

On the warth and the treacherous mud flats, I learnt to be still very early on and to track the tiny webbed feet after they left, before the tide washed all traces away.

The sighing song of the rushes waving in their dance to the variable and capricious wind, the rush of the tide, and the huge expanse of the ever-changing sky are the backdrop of my life. My dogs and I have never lived far from these shores and there have been many adventures along them.

My childish mind was lit by apparently endless fabulous sunsets and even now I witness both the dawn and the setting sun from where I live and I look west across the estuary. Being further inland, we have different birds, but it is nearly always the seagulls who are our weather vane, signalling the incoming fronts from the Atlantic, and warning of the autumn storms on the horizon when they seek inland shelter.

A couple of years later when I was back down on the warth with Benjie again, a seagull came swooping, swooping and I have always wondered whether he was my baby gull. Actually I am pretty sure he was.

CHAPTER 26

In Praise Of The Mighty Mouse

(Or More of Mice, Men and Cats)

This ubiquitous little mammal, more innocent and less worldly than their bigger cousins, is the subject of many children's books, of cruelty from cats and, sometimes, people endlessly depicted in harvest scenes and even our architecture, and they are famously celebrated by the Anglo-Dutch carver and sculptor Grinling Gibbons, who nearly always included a little mouse in his carvings somewhere. The humble mouse somehow seems to be an emblem of churches, country life, the animal kingdom and so much more....

It does seem quite extraordinary that such a tiny creature, in all its permutations, graces our books, galleries, houses, churches and even ducal monuments, as we recently discovered, when we saw a little harvest mouse at the very top of the 1st Duke of Beaufort's Monument in Great Badminton Church. The mouse peeks out with a very slightly smug and amused expression perched high up in the intricate flowers and leaves of the garland between the urn and the plinth at the top of

Gibbon's great monument to the Duke. There is an air of conspiracy too - as if the mouse is saying, "if you know where to look, you can see me", and "guess where I am?", which lends a childish enchantment to this great work of art.

I remember so many mice, especially their scurrying and tiny squeaks and endearing little faces and also the devastating destruction to our worldly fabric for their nests. Our cats would regularly bring in even more little mice from outside to cruelly taunt and play with them in the house but there was one particularly memorable occasion I recall. We had yet another pair of Burmese cats at the time, I think they were brothers, who once brought in a tiny mouse, which one of them carried upstairs in cruel triumph to my brother's bedroom.

The poor terrified little thing was still alive, for these two devils had a worse fate than death planned: my brother had a toy multi-storey car park for storing the toy cars. It happened to be empty that day, probably with the cars strewn elsewhere in some car crash fashion. The scheming cats put the poor little mouse into the multi-storey which had two exits, one at the top and one at the bottom.

Hearing a most terrible racket and somewhat blood-curdling sounds, I went to investigate, only to discover two cats guarding an exit each and the petrified little animal running up and down, up and down, the ramps of this toy car park. It was clearly becoming more frightened and exhausted by the moment.

Annoyed and somewhat upset by this James Bond-like torture, which had already gone on for some minutes, I am afraid I intervened and shouted at the cats who disappeared from the room post haste in high dudgeon at my interference. I think, but can't remember, that the mouse remained alive after its escape. But I nearly died of shock but at the intentional cunning and vicious cruelty of the cats.

A little harvest mouse, similarly captured for fun, had a luckier time of it. An enchanting small mouse, with that golden brown fur and sparkling brown eyes so characteristic of the type, was carried in and dropped in the hall, but the wee thing managed to run away and up the hall curtains. There it clung to the top of the curtain, looking pleadingly at my parents and me, clearly asking for sanctuary. This produced some consternation as how to best catch such a wily target and deposit him or her (we suspected her) outside in the relative safety of the field hedge. This was duly achieved with the aid of a ladder and a Tupperware box. The harvest mouse had been rewarded for her bravery and valour and we viewed the cat's sulks for the rest of the day with great amusement. And what is so fascinating in retrospect, is that the expression on that little murine face was not so very different to that of the Grinling Gibbon's harvest mouse.

Ellie and I had an interlude with a harvest mouse and curtains, here in Gloucestershire, which she seemed to find bewildering and I, amusing. One very early morning in the spring following her arrival with us, I was folding a rug in the garden room, when suddenly a little mouse ran out from under the sofa. I screamed, more with surprise than fear, and collapsed upon the sofa much to Ellie's bemusement. Christopher came down looking somewhat annoyed as he was preparing to leave in the early dawn for Kent and asked what all the noise was about; I told him and we thought no more of it.

That is, until the evening of the same day when Ellie and I were sitting upon the sofa watching television, when the said same mouse ran again across the room and, to our amazement, up the curtain and perched at the top of the curtain rail peering at us in a very similar manner to the earlier mice described. Ells looked at me in astonishment:

she clearly expected me to DO something. I didn't, I just looked at the mouse with a growing sense of déjà vu and it looked back at me for a while and then did something which even surprised me but Ellie even more.

The mouse resumed its rapid progress along the curtain pole, leapt from the crystal finial to behind the television, using its tail and forepaws to balance and steer itself, and then disappeared down the sideboard on which the television stands, never to be seen again. To this very day if you ask Ellie "Where's the mouse?" she looks at the exact spot where it vanished from sight.

Actually I think I did see it again, for there is a little harvest mouse which lives in our garden in the flower bed behind the pond, opposite the garden room. One day, early in the morning, it came out from the hedge and fixed me with its steady twinkling and somewhat cheeky gaze. That is as may be but I can say that it is clear to me that these little mice certainly like to steal the show and are not going to call it curtains.

CHAPTER 27

PIGLETS

Down the road from us at The Old Parsonage, there was a farmer who particularly enjoyed breeding pigs. Clearly his delight and fancy, he was often seen delicately stroking the back of a sow, he was about to slaughter or take to market, in the manner of a P.G. Wodehouse character, which always seemed somewhat perverse to our young minds.

On one occasion, a large number of piglets, which were complete rascals, escaped. They ran up the road, into our garden, and then into the orchard where the trees were laden with ripe apples of many different varieties: Cox's apple pippin, Burnet, Bramleys and other old English varieties. The weight of the crop was bending the old lichened branches near to the ground and the fruit was beginning to fall. If you were still and quiet, you could hear the thud of the apples as they hit the ground, as if in a perpetual action replay of the origin of the theory of gravity.

Squeak, squeak, squeak, they said as they ran through hither and thither; run, run, run, they ran on their little trotters; scoff, scoff, scoff, they said to the apples with the joy of piggy victory (food) and a truly excited look in their little eyes. By the time all twenty one piglets had been finally rounded up, every single one had an apple in its mouth.

As my mother said "Ready for the oven". I thought that rather cruel, but the sight was worthy of eing recorded for perpetuity. Complete chaos, absolute abandon, and totally hilarious.

The piglets were duly returned to the farmer and the story became a bit of legend; my father said that we could have had nice bit of pork. I haven't liked pork ever since.

CHAPTER 28

Bat Rescue

The day that we acquired our first bat, was to our childish minds somewhat unusual, and an event of much hilarity and some wonder, if not outright disconcerting only because we were used to the vagaries of our family and domestic animals.

We were going to Mass one Sunday morning around 8.30 am in the morning, (I can't remember exactly what time of year, it was fairly light but a little grey so possibly the Spring). Suddenly my Mother gave a little shriek and pulled the car over exclaiming, "There's a little bat in the road and it's all alone!"

She got out of the car and went over to the poor thing which was very small and looked, from my now adult perspective, dehydrated, certainly frightened. My brother and I exchanged looks – what on earth was she going to do now?

My Mother bent over and gently picked up the bat from its position in the middle of the road, where it had been so vulnerable and gently brought it back to the car, where she proceeded to place the tiny bat on the back seat between our shoulders.

We were all used to rescuing animals, insects, and forlorn creatures of any kind that could be kindly handled but this a new one to both of us! The remainder of the drive to church took on a rather surreal quality with the bat clinging onto the fabric of the back seat, we hardly dared move for fear of disturbing it.

During church I gingerly anticipated the drive home with some concern – what if the bat were to move or try to fly? What if it had moved while we were in church? When we returned to the car, I heaved a sigh of relief, the bat had not moved from its original position. Now for 15 minutes to get it home safely...

Having held our collective breath virtually all the way home, we finally pulled up in the stableyard and the little bat was duly and ceremoniously removed from the car, and placed with easy access to water in the hayloft in the coach-house.

In the ensuing months something of a miracle occurred, and I could never work it out...our one little bat somehow turned into two, then three and then before we knew it had a small colony. I can only assume that the little bat, who was not so little by the autumn, had communicated by sonar to other little bats...

Summer evenings at warm dusk, when the burbling of the swallows had subsided and the wisteria released the last of its spicy evening fragrance, and the nightingale had put to rest its heavenly song, and the night would be alive with fireflies and the sound of grasshoppers, the air filled with fluttering wings of a different sort. Only young ears could hear the tiny high pitched squeals of those winged mice as they ventured out on their nightly expeditions.

And all from one tiny little abandoned bat? The natural world moves in seemingly miraculous ways at times.

CHAPTER 29

Ferrets And Pubs

My brother came home one day with two ferrets, a grin lighting up his face.

"You are lucky they are not polecat ferrets," he said, thrusting one of these unknown creatures, with a fearsome reputation and not known for their sweet smell, into my arms.

"What an earth do we do with them?"

I do not remember how we came to the conclusion that one of the out-houses, not the swallow villa, should be the place, and I am unclear as to where to their temporary accommodation was, but they were to prove their true worth to our menagerie.

The early months fell to my brother to manage and negotiate them (after all they were his pets) so I had little to do with the weasonbily recognisable, but stoatily different creatures. They were taken out and put down rabbit holes, fed on I don't know what, slept I don't know where, except I recall a cage...and then one day, they disappeared.

Two of them, two little white (on a good day) balls of mischief, terrorists to the cats, disdain for the dogs, interest, sometimes for the chickens, and amusement (for us). And then suddenly one day they were gone…

The sadness was palpable – these little mavericks had tricked, laughed, endeared themselves to us all by just being themselves – difficult, capricious, funny, persistent, and then, without warning, having bitten their way out of their (palatial) cage they had vanished…leaving a hole in our hearts rather larger than the rabbit holes to which they were so accustomed…

"Gone to mate with the polecats.." was the received wisdom but nothing assuaged the empty cage, until one day we received an enlightening telephone call (perhaps enlightening in the sense that it might also explain my brother's unauthorised absences).

"I don't know if it is, but I think your ferret is in the car park," pronounced the local pub landlord; I had taken the telephone call chanting "Pilning 2236", which now seems a most antiquated number. This was a line which was frequently swamped by the incoming tide from the river for hours at a time, which was to prove to me massively frustrating in later years, when I was experiencing the typical teenage angst of waiting for the latest crush to ring.

But this was better news than any prospective boyfriend, the ferret, our ferret was alive after all those months.

"Only one, I'm afraid".

No matter, one is better than none. My brother was duly dispatched to collect the fugitive who had apparently accosted a visitor in said car park and turned himself in (perhaps living wild is not all it is made out to be).

Clearly not, as once restored to civilised domicile, Ferrety, as he was known from then on, made a bee line for his cage and let himself in with what was clearly great relief and gratitude, and awaited his dish of cat food with eager eyes and even greater appetite. Having taken splendidly of the offered repast, he curled himself up in little ball, nose to tail, feet tucked in and fell deeply asleep.

I, the disdainful female teenager who naturally hated ferrets and all they represented, fell in love, all over again…Ferrety became one of the most extraordinary creatures I have ever known.

My brother went away to school, so I assumed responsibility for Ferrety, as well as the chickens, feeding the dogs and the cats and any passing animal tramp (or even tramp). To let Ferrety out of his cage at dinner time and race through the courtyard into the kitchen to eat his dinner alongside the cats and dogs (the rule was they ate together) was as joyous as anything I have ever seen; notwithstanding some vicious looks (cats), a few low growls (dogs) he persevered and night by night and silently he made friends.

He never hunted again but instead confined himself to wandering around the garden, often at my heels, or wrapped around my neck (I learnt to be careful, though, after my washing my hair he was inclined to be amorous with my ears, just a little nip and tuck (which hurt)), smiling, learning to walk on the lead, stay down a leg, sunbathe in days that seemed so glorious and warm and endless, until most of the time he was my companion out of his cage along with all the rest… revising Greek and Latin, maths, French, geography, biology, history (spectacular fail), trips to school hidden in my Barbour, he was my co-student, curled up in the sun, warm content and if ferrets purr (why polecat?) he purred and stretched and purred…

Truly that little ferret was a master of how to live life well and lovingly. Adventurer, flirt, crusader, antagonist, beauty queen...he became quite the star.

CHAPTER 30

Goofy Guinea Fowl And Mountaineering Pheasants

I woke early one morning on a weekend visit home and drew back the curtains to discover my mother's latest embellishment to the grounds. I was most surprised to see some incredibly ugly birds running all over the lawn, and not only did they look ghastly, but they were making the most infernal din, as they chased each other across the lawns.

"God, Mummy what have you done now?"

The following morning when I rose and walked to the window to look at the day, there were these birds again, mad, bad and beautiful, behaving as only guinea fowl do scattering, twittering, quarrelling, and generally getting along. They, I thought, as I watched them with fascination, and not a little admiration, invading their new territory, are a real lesson to us all in their bickering, snickering, sulking and love. They moved as a flock, like a cloud, across the lawn and the sound of their feathers was like the sound of angels passing as if each wing was being lifted by the breeze of movement in some celestial purr.

You will have realised by now that it did not take long to succumb to Guinea Fowl Fever; far from continuing to consider them so ugly, I began to list all the little physical foibles which made them so beautiful; from scattering like a breeze, the delicate white and grey patterning on their feathers and the jaunty dash of red highlighting their jowls and head. Their eager, eager little mutterings of excitement, especially when they were engaged in their latest nefarious adventure whether it was to raid the precious and coveted daffodils of the village (many heated telephone calls ensued), annoying the holiday makers at the caravan site next door by tapping on their doors with their beaks and then running away, or merely refusing to be rounded up for bed by us as they were very much targeted by our resident foxes.

Eventually I persuaded the fowl to join the other fowl in the coach house for the night time so they were safe. They did not like being told to go to bed, so the evening ritual of laying the birds to rest became a nightmare. "Go away", they spluttered, tripping over each other in their eagerness to demonstrate that they did not wish under any circumstances to be confined with the other fowl. "Go away, away..." and they squawked "Go back, go back", the eternal cry of the guinea fowl. "Go back".

I became extremely fond of these daffy charming birds and whilst still pretty untamed, they would flock up to see me if I sat quietly under the three hundred year old green oak where they would perch randomly up branches or on the ground, but always as the gang, hating to be alone. They embodied the true meaning of the words "birds of a feather flock together."

They also liked to stray out onto the road, creating a huge hazard for any visitors and sadly sometimes being injured or worse as a result. Around here, pheasants are a constant danger to motorists, simply through their refusal to acknowledge until the very last moment that they have wings, and to take to the air after a dangerous swerve by an oncoming car. A game of chicken which sometimes does not end as happily as an airborne bird living to survive another day's shoot.

One of the roads near here C and I have nicknamed "The Pheasant Run" due to the large number of kamikaze birds which litter the roads after the "rush hour". Many of the birds, sadly, do succeed in their quest for immortality, but the roads are remarkably clear of the fallen within an hour or two of 8 o'clock.

There is, I believe, a curious anomaly in English agricultural law, that deems it illegal to pick up any bird that you have directly hit with your motor car, though it is perfectly okay for the following driver to do so, innocent, I suppose, of the crime of accidentally slaughtering the bird.

Recently, we witnessed a bizarre and amusing sight of three pheasants up on the apex of a neighbour's roof, to the disgust of the local jackdaws and crows, who were hissing their disapproval of these foreign immigrants silhouetted against the early morning sky.

Looking for all the world like walking weather vanes, two of them elegantly strutted along the apex towards a further peak, crowned by an even more daring feathered friend; they once again exhibited their disinclination to use their wings, preferring to herringbone the route up to the top, only to disappear from sight as soon as the inevitable iPhone camera appeared to witness this unusual roof top embellishment.

When at Aylesmore we were always amused to see we had a dashing, if stupid, visitor to the hens in the autumn. A pheasant would without fail a wooing of the pullets come, as if fresh from sunny foreign climes with his gorgeous get up and exotic colours. He never failed to get the girls in a twitter, and though we were unaware of any success on his part,

I have heard it is possible for there to be offspring between pheasants and hens and there is a lot of discussion about such endeavours on the internet. Whatever it was amusing, if a little sad, to watch this elegant gentleman stroll in from the fields each day to pay court, though we wondered what on earth was wrong with his pheasant hens.

These feathered friends, when not coming to a sticky end, do seem to provide an endless source of entertainment. When we were growing up amidst all this animal activity, we would remark often: "Who needs television when you can enjoy a moving landscape?"

CHAPTER 31

Chicks, Chickens, Carnage And Miraculous Survival

My landlady at Aylesmore in the Forest had an amusing story about chickens: her father, who had been a General stationed in Germany, once owned a highly prized collection of leghorn chickens. The family had gone out to a drinks party at the nearby embassy and came back to discover to their great consternation that they could find no chickens in the grounds. They looked everywhere, but found not a single fowl…

Defeated and downcast by their unfruitful search, they went into the dining room and found, perched upon each of the dining chairs, a charming sight: a chicken sitting up at the table ready for lunch. Looking highly expectant they called lunch…lunch, lunch…lunch…

Shooing them outside to some great disappointment and disapproval from said assembled feathered invaders, there was a good deal of clearing up in the dining room before lunch was to be had!

Over the years we had quite a collection of chickens of various different types and characters at the Old Parsonage, including the war-like Earl of Essex who sported the most charming Elizabeth ruff, especially when he was engaged in his favourite sport which was beating up the other cockerels, however young or brave. He always, it has to conceded, came off best in these sparring matches.

The hen who particularly stands out in memory is Ermentrude, named after the cow on *The Magic Roundabout* for reasons I cannot recall. Ermentrude was the greatest and toughest bird I have ever met; feisty, vain, beautiful and bossy. I also nicknamed her the Joan Collins of the poultry world, meaning no disrespect to our thespian Dame. Ermentrude was a beautiful honey coloured personage who would bathe endlessly in dirt baths, preen and sun herself, conduct herself with great dignity and keep all the rest of the flock well in check. Should anyone have chicks, she would take it upon herself to make sure the mothers attended to their duties of rearing their young and in every way ruled the roost.

Then disaster of the vulpine variety struck: the local foxes had become more and more vicious and daring in their attacks upon the hens and the guineas, which was not helped by the fact that the Berkeley did not hunt in the immediate area due to the proximity of the motorway and an unfortunate incident some years previously when the fox led the hounds off the cliff at Aust. My parents came back one time to find the entire flock eradicated, not even at night but with great audaciousness in the broad light of day. Every single one beheaded by a gang of foxes and not one left at all. The scene was one of utter carnage, a sea of tossed and discarded feathers and corpses.

Our whole family was devastated and we vowed never to have chickens again, I personally spent weeks feeling bruised emotionally by this incident, even though aware that it was only a natural, if cruel, event.

However, our vow did not come to pass, as some weeks later a rather dishevelled and weary chicken turned up at the back door. Recognisable only by her distinctive honey blond colour and her peerless squawk, we realised that somehow Ermentrude had risen like a phoenix from the ashes of disaster. Once she was nursed back to health, we duly acquired some assembled company for her and she once again rose to the challenge of leading the troupe.

One thing was different though, whenever the grounds were to be deserted for more than an hour or we were to be out for the day, the hens were confined to their (large) quarters much to Ermentrude's vociferous objections. There was, though, no repeat of the sly and wicked slaughter that had marred our hen husbandry to date.

CHAPTER 32

Of Mice, Churches, Christening Cake And Christmas Stockings

To be as poor as church mice is an old fashioned way of describing being dead broke particularly when newly married and seems to be well rooted in the English language and collective psychology.

However, in my experience, the church mice I have met seem to live in some splendour compared with the rest of us and have the destructive ability, at times, of a mighty tornado. Not only do they possess the ability to chew through electricity wires and so pose a significant threat to human safety (I once heard of a nun, at one of my schools, who was unfortunate enough to die in the bath after being electrocuted by open wires chewed by mice), but they can disable the mighty organ in the church and cut off the power supply to the house!

However, to the more charming side of the murine character: I once met a diminutive church mouse who lived under a table, behind a curtain shielding the Great West door, in an old church where communion was regularly held so there always flowers on the altar. One day we solved the problem of how it survived for food, when we found the flower vase knocked over and the wheat stalks clearly pilloried and squirreled away in some tiny undetectable larder. We also had a similar episode at one of our churches here at Badminton. We could not work out who the vandal was destroying the flowers from week to week until the seeds were clearly visible strewn across the floor. Not only food, but easy access to water outside, and occasional fine music…sounds an ideal existence to me.

Less pleasing was my run in with our murine companions some nine weeks after my daughter was born and my ex-husband and I were generally living in different parts of the coach-house which we rented in the Forest of Dean. One week before my daughter's christening, I went to the sitting room the other side of the house to prepare it for her christening party and started to plump up the cushions on the sofa. To my horror, when I pulled out one of the main cushions in the body of the sofa, I found a very comfortable little house for what was clearly a large family of mice, of whom I had hitherto been entirely and blissfully unaware. Exploring various areas of the house further, it became apparent that in the absence of my attention, the mouse population had exploded and we had a serious vermin problem.

Reluctantly, we came to the conclusion that we had no choice but to put down traps; I hate poison, traps are no better, but we were at a loss as to the best way of proceeding. This decision led to some hysteria, both fear and laughter in the ensuing days, on the Nanny's behalf especially, as she described how she had witnessed in the night, the clever mice using the traps as a springboard to get at whatever delicacy we had placed there, chocolate or cheese, and then run away with their prize.

I was completely defeated by the diminutive conquerors when, the day before M's Christening, I found the very grand wedding cake tier thoroughly nibbled and elegantly attacked by our tenants. Waitrose was visited for alternative cake, though the original still went on show, a Jemima Puddleduck cake presided over the Christening tea and the subsequent party that evening.

Oddly enough, soon after the noisy party our tenants departed and there was no more evidence of their tenure, so Beatrix Potter had the last say.

And so to the Christmas chocolate. While living at Olveston, when M was six, I had hidden away in the cupboard under the stairs, the offerings for her stocking including, perhaps foolishly, the traditional golden chocolate money. Putting away yet another paltry gift, I was perturbed to discover that someone had clearly been helping themselves to the hoarded riches, and that someone possessed very tiny teeth. Exploring further I came upon evidence of the thief and unkindly put down some poison as I was, unjustly, cross. A few days later I decided to check the tray: I was so ashamed when I found it.

Carefully laid across the poison were strips of material clearly meaning "Keep Off". I pulled back the pile of material under the stairs and there found the most enchanting supplicant. A mouse stood her ground with her very young babies behind her and stared me down, but with pleading eyes.

I sighed and said "OK you win, but we need to make a bargain – if I take away the tray, you stop eating the chocolate". I removed the tray, the chocolate was left alone and come the Spring, there was no new trace of mouse in that cupboard. Once again I had testament to the extraordinary bravery of these exquisitely small destructive and delightful creatures. I bow to the mighty mouse.

CHAPTER 33

Christmas Trees And Christmas Presents And Robins

When I was a little girl and we were still in Bristol, my parents had quite a collection of Christmas ornaments: the traditional glass balls, which the cats enjoyed shattering at any given opportunity, the angel, the multi coloured lights which inevitably fused every single Christmas, as lights always did in those days, and the other old family favourites, which are so different to the decorations favoured nowadays. As was traditional then, the Christmas tree did not go up until just before Christmas Day, right towards the end of Advent, and so was an event of huge excitement. My favourite decoration was a little, extremely realistic, Robin, which was always placed near the top of the tree, as near to the angel as possible.

Two days before Christmas, just after we had finished decorating the tree, we all dressed up and went to the local Carol Service which was very exciting, for we were allowed to stay up late and would have hot chocolate on our return home.

Well that was the plan anyway. The Carol Service was lovely, we did stay up late, we did have hot chocolate and opened the penultimate window on the advent calendar, but we also came home to a scene of utter carnage in the sitting room where the Christmas tree was.

Well, had been, to be more precise! The tree was lying down, the lights inevitably fused, the shiny glass baubles were for the most part shattered, the angel lying in a most undignified position, but worst of all was the Robin…

Which evidently in our absence had been murdered.

Poor wee thing, lying there with its bloodied breast and badly mangle feathers, looking even more real in this scene than it had done when it was perched high up on the tree.

My parents had the grace to laugh. Apparently the cats had been stalking the Robin since the tree was decorated, and clearly had taken the opportunity, in our absence, to commit a bit of bird genocide, by climbing the tree and killing the not so real bird! Imagine their shock, when the tree came down, the baubles shattered, the lights went out…

The Robin actually lived to grace the restored tree having been mauled, not killed, the glass ornaments were not so resilient, the angel was restored back on high, and the lights relit, and though the tree was somewhat bare, order was restored.

We found, too, two very frightened, though probably unrepentant, cats who had met their match at the top of the Christmas tree.

Many years later, it was the bottom of the Christmas tree that came under attack, this time by Robbie. At the time I was living south of Bristol and I had a shop in Clifton. Some days previously I had put up the tree with M, decorated with wonderful trimmings as I had been a designer and importer of them, and so had my pick. There were gold apples and pears, musical instruments, strutting peacocks, sparkling white doves, white twinkling lights and Crabtree and Evelyn's wonderful Noel spray, which has become a Christmas hallmark for our families.

I had already surreptitiously wrapped a huge number of presents for M, being a single mother, by then, I always overcompensated, and placed them under tree glittering in their golden wrapping and extravagant bows. I was so proud of the way everything looked, as it was our first Christmas on our own, though we were having everyone for Christmas Day, including her father.

So, M and I had been attending her School Carol Concert the day before breaking up from school for the holidays, and we were, therefore, a little later than normal getting home. Normally I would have had Robbie with me but I had taken him home earlier as I did not want to leave him in a cold car on a starry December night. (Bristol churches are not as welcoming to addition of dogs or other animals to their services as we are round here.)

I did not notice anything wrong when I came home and only glanced into the sitting room before feeding child and dog, and putting a tired child to bed. It was only when I went into the sitting room after that and sat down that I noticed what was amiss.

The presents under the tree had been unwrapped, fairly systematically by the look of things as the paper was not too tattered and the bows still intact. The contents were pristine, even those of the ones for the…

"Robbie!!!!"

I had wondered why he had a slightly shifty look on his face when we came home and was rather reluctant to join me in the sitting room.

Guilty as hell was the look written all over his face, and I said "Well it could not have anyone else could it Robs?"

It was hard to be cross, it all looked so neat, as if some careful child had opened the gifts and not some doggy teeth and paws. In fact it was so hard to be cross, that I laughed and the look of relief on Robbie's face was palpable. Then his look said, "That'll teach you not to leave me with them under the tree and come back late".

When I rewrapped all the presents, I was careful not to put them in the usual place until Christmas, lest the temptation prove too much again.

In later years Robbie eyed up the presents under the tree but the trick was never repeated, at least until Christmas Day, when he finally got to open his!

CHAPTER 34

It All Depends On Your Point Of View

Once, going round Bristol Zoo I said to my late father,

"Are we on the outside looking in, or are they on the inside looking out?"

In other words, who is entertaining whom?

This abstruse idea gained some credence during the Foot and Mouth outbreak in the spring and summer of 2001. Our country side being ravaged by this appalling disease with more than 2,000 farms affected, and even more appallingly contained and managed. The general election was postponed a month, many sporting fixtures abandoned and the countryside was virtually in shut down. Smouldering bonfires testified to the terrible slaughter amounting to some 10 million sheep and cattle. From our house in Herefordshire, we could see the funeral pyres.

By the time the disease was contained in October 2001, (or the virus had come to its natural conclusion as they can), the emergency had cost Britain £8 billion in agricultural and countryside tourism lost revenue and expenses. Moreover, the outbreak was traced to infected meat that had been illegally imported.

Anyway, the Zoo was closed and this, apparently led to a sustained bout of depression in the Seal and Penguin House; used to the daily entertainment of either entertaining their visitors or being entertained by them, it is not clear, they fell into a decline of sorts, which was only alleviated by the afternoons being dedicated to sporting pastimes.

Given there was less for the zoo keepers to do given the zoo's closure, there was more time to play with the animals. So the Penguin Housekeepers apparently took to playing with their charges during the afternoon, thereby raising the spirits of all concerned, until the Zoo re-opened in early April of that year.

I tell this story as evidence, weak maybe, and at the risk of anthropomorphism that animals seem to benefit as much from their association with us as we with them, with the caveat of course that they are well treated.

The phrase, "It's a dog's life" always seems to be somewhat ironic to me, particularly if said dog is the loved and pampered pooch that most of our domestic pets appear to be. Oh, for a dog's life, we might all think from time to time. But we restless mortals would probably not appreciate it in the same grateful and giving manner even if Ellie does proclaim she's bored on a fairly regular basis.

CHAPTER 35

Another Helicoptering Dog

Robbie, as I have mentioned, was capable of leaping up huge heights to the point that we nicknamed it "Helicoptering", as it would appear to the onlooker that he was jumping from a standing start. Benjie had already opened my eyes to the incredible athletic prowess of the lurcher, especially the males, on a number of occasions, but none more so, than when he displayed great irritation at human indecisiveness and took matters into his own paws.

Some years previously, my father and I were walking in the Brecon Beacons on a trail we called the Waterfall Walk, and it was absolutely tipping with rain, so much so that we might have been standing under a waterfall anyway. As is often the way, these managed paths are not always dog friendly in that the stiles are high and sealed with chicken wire at the base. Additionally, they are roughly four and a half feet high, and that two storey construction is so irritating to the dog owner. There is no option, except to lift your companion over.

This is fine if you have a dog of a manageable size and weight, not a large lurcher of Benjie's proportions. My uncle, who was walking with us, pressed on with my brother, whilst my father and I stood by the very tall stile. With a sea of mud around it, we were wondering whether this was the end of the walk for us, as it was evidently going to be very difficult to lift Benjie over.

Standing between us, Benjie's head moved from side to side with increasing exasperation and a dogged look of annoyance as he listened to our somewhat fractious and worried debate.

"We can't lift him over that…it's too high…I do wish they would mark these stiles on the map…" and so on. His eyes signalled, quite clearly, that he considered us utterly incompetent and that our hesitation was vexing, to say the least, for a lurcher who was enjoying the new walk so much, and with a final look that indicated something along the lines of, "…. You dithering humans, I'm making my own way," he leapt from his standing position without any warning whatsoever. He cleared the stile landing somewhat messily but safely in the mud the other side, and stood there looking at us with a smirk on his face, "Idiots". We both climbed the stile, somewhat chastised by his expression, and proceeded on.

On that same walk we had a curious incident, approaching the area where the waterfall was marked on the map. It was in a very old woodland area with many gnarled and bent old trees, and with the mist from the damp weather swirling around, we could only locate a stream but not the exact position of the waterfall. This was of vital importance as it was the destination of the excursion as this cascade has a legendary phenomenon. If you go along a very narrow and dangerously slippery ledge from the main trail, you can reach a place where you can actually stand behind the falls themselves.

We could hear the thunder of the water, but we could not see where the path was, even my uncle was unclear and he was a quite famous explorer! Suddenly out of the deepening mist came a curious figure, an old man, small, bent but still extremely agile, judging by his bearing and the jaunty swinging of his stick. When my father hailed him he turned and leapt the not so inconsiderable stream, which was littered with mossy granite rocks, and quite noisy.

"You'll be wanting that way," he said with a knowing look in his bright blue eyes, plainly he was used to clueless walkers, and brandishing his hazel wood stick he pointed out the path to our target destination.

We thanked him and turned to pursue our final ascent with renewed purpose. But here was the odd thing, Daddy and I turned within seconds of completing the exchange, only to find there was no one there, just the wood and the mist and the sound of water. We looked at each other and shrugged, pretending we thought our merry fellow had been swallowed up in the mist.

Later that day, my uncle and father discussed this, and my uncle said that in his experience it is odd that the weary traveller will often come upon a guide who disappears as fast as they appear. Curious maybe. My uncle, many years later, related to me one Christmas how, when his climbing companion in the Andes had injured his leg, and he had had to get him down the mountain as quickly as possible. In his fear and, perhaps desperation, he felt, and even saw, clearly a guide of some kind who had disappeared as soon as they reached a place of relative safety. It would be easy to dismiss such stories, if they were not relatively commonplace amongst travellers, explorers and others in extreme conditions.

On the way back, we left Benjie to his own devices to negotiate the tricky stile, and contemplated how often we are blind, both to the intelligence and resourcefulness of our animal companions, to say nothing of their ability to express their disdain of our problem solving abilities without saying a single word!

CHAPTER 36

River, Water, and Nine Lives

Visits to the river have not always been so quiet and uneventful in terms of drama. Benjie loved the beach down at Aust, but one of his early encounters was nearly a total disaster.

There is a convenient causeway created by the electricity company, so that the service vehicles can reach the national grid pylons which straddle the river. It is so sad that these punctuate the sky with their ugly and man-made lines, but interestingly as a child I actually was able to blot them out in my mind's eye when surveying the local landscape. It is only since I moved here to an area where we do not, thankfully, have the pylons, that I began to notice them again. However, the causeway is a most useful approach to the beach, especially at high tide when the water laps up against the edge, and indeed at the neap tides and equinoctial tides, it is completely covered.

And this is precisely why our little walk with our newly acquired puppy nearly ended in disaster. My mother and I were taking Benjie down to the beach probably for the first time, certainly he was unfamiliar with its dangers. All went well and we were enjoying the feeling of being above the water and able to enjoy the sensation of being amongst it until....

At the end of the causeway, there is a slope down to the beach and, on the left hand, side a steep wall which gives on to the marsh mud flats. At least, that is what it looks like in a normal tide but on this particular day the water was much higher than we had been expecting. What happened next was even more unexpected. Reaching the end of the road which was dictated by the tide as the water was lapping at the top of the wall on the left hand side and surging up the causeway slope. We turned for home.

At least the humans did, but the puppy had other ideas. Clearly thinking this would great fun, and without any sense of danger, he launched himself off the causeway towards the water and the outgoing tide.

It was one of those moments which seem to stretch out for ever.

As I started to turn around for home, expecting Benjie to turn too, out of the corner of my eye I perceived a sudden scooting clumsy puppy lurch and launch over the edge. Without even pausing to think I dropped straight to the ground, and just as he was about to impact with the murky river, I grabbed his collar which luckily being a lurcher collar was wide.

Even so I was fearful that the little mite might slide out of the collar, so as quickly as I could, I grabbed his bottom, most unceremoniously, and picked him back to the concrete shore.

A little wet but completely unabashed, he shook himself off and ambled off towards home with a reproachful look at me, as if to say that I had injured his pride! A sign of things to come as he was a cussed animal at times but we were so relieved. It had never crossed our minds that we might have a river dog but he always loved the place so there were no nasty lingering memories of his first brush with disaster, though I was a great deal more wary thereafter.

Many years later at Aylesmore, Robbie's close shave with the well shaft reminded me of this incident and we often joked that he must have been a cat in a previous life, due to his determination to use up all of his allotted nine lives. What was so peculiar was that it echoed this earlier incident with Benjie: in both cases, my reflexes were faster than my conscious thought, otherwise both of my darling dogs might have been lost at much younger ages than they enjoyed.

CHAPTER 37

KINGFISHER

A flash of bright cerulean blue, a sliver of orange, a dart of wings, all along the other side of the riverbank. By a small cliff, under the verdant haze of the overhanging trees, there were these tiny explosions of colour and speed.

"Robs, still," I whispered, "Very quiet now, down".

My heart sounded incredibly loud as Rob and I stayed stock still, I having dropped to a crouching position amongst the grasses weaving along the bank. It was very early in the morning and we were unlikely to be disturbed, the mist was rising from the little river and the sky was clear promising a bright day, the rising sun slanting through the willows and turning the river water to bright pools of light being dashed into thousands of liquid diamonds against the rock bed.

And this in the middle of a suburb in Bristol.

Blaise Castle is a wonderful 18th Century Estate in Henbury in Bristol, not far from our home in Bristol or my school, Badminton. I had known it well as a child, but not to the extent I was learning about its hidden life and wonders as an adult. I walked there with Robbie every day for a while, while M was also at Badminton and I had the shop in Clifton. Every morning we would go there whatever the weather and I came to know the other regular walkers quite well. They were a great solace to me at a time in my life when I was very lonely indeed, except for my beloved daughter and Robbie.

There is a wonderful gorge, along which in the 18th Century, the Harfords, who owned Blaise at the time, had created a fantastical winding drive. This was created in the true sublime tradition, so influenced by The Grand Tour, and the subsequent infusion of all things Italian and alpine into our culture. (My great-great-great grandfather also had lurchers as carriage dogs, and on the Grand Tour, he had little leather bootees made for them to protect their delicate pads from the stony roads of the Alps, so different to our grassy pastures and leafy woodland landscapes.)

Here we both were, still as the grasses around us, just the sound of the river, distant traffic, and intense breathing, as we observed the other side of the river until that Kingfisher was gone in a bright burst of colour.

After that we would visit that area of the walk as often as possible, learning more and more about how the Kingfishers lived, and ever more discerning in our observations. Robs never let me down by moving suddenly, and so he helped enable one of the most extraordinary experiences I have ever had the privilege to enjoy.

One morning very similar to that first time, Robbie and I were quietly watching when…

Suddenly, darting across the little sparkling waters came the blue and orange bird with its distinctive yellow beak and alighted on a willow branch, no more than six foot from us. The Kingfisher cocked a sideways glance at us, before settling its gaze to the little rippling current below his branch, where the green leaves grazed the water.

Had we been still before, we now held our breath, and also turned our gaze in the same direction toward the water though sneakily, in my case, casting a shifty glance to my right to where the stunning bird still sat motionless on the slender branch. It was like companionship...almost as if we were being let into this avian paradise and the whole world took on a translucency and brightness. Whether through sheer exultation or surprise, I shall never know. I do not remember how long we three remained in what felt like silent communion in the early dawn light.

Without warning, he dived and then was gone.

Almost a sense of bereavement assailed me, as the curtain to another world had been abruptly shut. We moved gently off to our normal meandering but swift walk, I for one wondering at the miracle that seemed to have taken place.

Perhaps my experience is not unique but those few minutes side by side with the elusive Kingfisher changed my world for ever; it still feels like a miracle and was a truly mystical and transforming experience. Just writing this makes me aware how lucky I have been to have walked upon the seashore and seen the birds and wild places that I have seen and met my little seagull amongst others. The crowning glory of my avian experiences, though, still has to be that moment of complete stillness by the Kingfisher.

CHAPTER 38

Equine Exploits

Not mine I fear, and I feel very deeply my lack of equestrian expertise, as I come from a family of excellent horsemen on both sides, and my life has been enriched by so many stories of the heroes, champions, scallywags and rascals of the equine world.

In my defence I had quite a bad accident aged 16 or so with a horse, who also managed to dislodge my father, who was a superlative rider, with a huge amount of experience.

Though that was the end of my riding career, it most certainly was not the end of my life-long love affair with these magnificent creatures, and we have ample opportunity here to appreciate them in all their different varieties, on a nearly hourly basis.

"Tracks and roads, tracks and roads", is the echo of the racehorses' hooves, as they trot up to the gallops at first light; the sound of the hunting horn shimmies around the surrounding coombes and valleys; the rumble of the horse boxes, as they troop to pony club in the summer holidays; the traffic on the road and the palpable local excitement at the Horse Trials held annually, (unless foot and mouth or inclement weather dictate) at nearby Badminton, one of the world's largest sporting events; the happy birth of a beautiful foal from a gold medal dressage winner....I could go on. And that is just here, not where I grew up with the Berkeley and the stories regaled over supper, in a warm kitchen of chilly days out hunting on Exmoor, even colder stories of death on the road. There were happier recollections of winning championships and horses getting you home safely even if you are half asleep in the freezing cold night of the Canadian backwoods.

So many stories and the best of them to be recalled with the best of life, by the roaring fire, when the equines are safely fed and bedded down for the night. Our extraordinary bond with not only our dogs but our horses is the legacy of our land. We forgot that horses were only recently supplanted as a means of transportation some 150 years ago with the advent of the motor car, a mere comma in the history of civilisation.

My first encounter with this animal group was before I was born, with a pony called Honey, who was a veritable minx and the clever clogs of the wild Welsh pony world. She was my Grandmother's little pony, and apparently, during the war they were a common local sight as they traversed the moor, doing Gamma's (her preference for Grannie) war work, whatever that was, the precise definition of which has disappeared in the shrouds of time.

Honey was a character. She was an escape artist, wild pony, avid huntress, loyal companion, clever protector and much more. The story that particularly enchanted me as a child was her savvy protection of my father when returning from hunting late one winter's afternoon.

The sun was setting fast after a long day on the moor and the shadows of the twilight were conspiring to lull my father into a soporific reverie on Honey's back, as the rocking motion of her quiet but determined lollop home soothed him into a near sleep.

Until Honey came to an abrupt halt;."Go on, Honey, go on," my father urged her sleepily, "Move on there, girl."

She didn't budge, and reluctantly her charge opened his eyes to find that they were standing stock still on one of those narrow paths which lace the valleys up from the coombe to the top of the moor where Woolhanger, my grandparent's house stood. Honey knew her way home well, but she had come to a full stop, not through lack of navigational or homing skills but for a better and very solid reason.

To their right in the growing dusk rose the steep hill. To their left, it tumbled away down into a deep ravine, ahead right across their path was a large solid fallen tree, which if she had jumped with my father in his somnolent state, would have probably dislodged him. He might have fallen, tumbling down into the valley far below and to an uncertain fate.

As it was, confident that my father had finally opened his eyes due to her stubborn refusal to move, and that he was aware of the situation, she jumped the tree, which was no mean feat in itself, and continued to trundle home as if nothing had happened.

And it hadn't but my father often wondered if it could have had, and remained most grateful to Honey for her thoughtfulness and common sense.

She, too, was inclined to be an escapee, and like Robbie had it down to a fine art. Every spring after the hunting season had finished, she would be turned out to grass and all would be well for a few weeks. Until the inevitable cry of "Honey's left again" would echo through the stable yard at Woolhanger, my grandparent's house.

You see, Honey liked to spend her summers on the moor with the wild ponies and nothing, apparently, would deter her. One late spring, my uncle and my father were determined to fence Honey in and they spent a day constructing a new and immensely high fence, partially composed of brush. Honey watched the proceedings with great interest, and in retrospect, some amusement. When they stood back at the end of the day to admire their handiwork, she looked at them with a horsey grin, and retreated to the opposite end of the field from where she proceeded to gallop full pelt, clear the new enclosure and was away to the hills again.

She returned, as was her wont, in time to be walked out for the hunting season looking very shaggy and quite wild but very fit and full of herself. Honey is long since dead, she died shortly after I was born, but she always seems immortal to me, because of the stories that the family told of her exploits, her courage, her cunning, cleverness and thoughtfulness. It is through the acquaintance of these animals, not only in life, but in family myths and tales, that I have come to appreciate the various and deep qualities of so many animals. I wished to set down some of these tales, which seem magical and celebratory to me, for posterity.

CHAPTER 39

Highgrove Hens And A Heckled Dog

Unfortunate Forays Into Fowl Housekeeping

My most recent excursion into hen husbandry ended not in disaster, but in sadness. It began with great high spirits, as the Duchy Home Farm at Highgrove were off-loading some of their older layers, and so we hot-footed it up there to spend a little time chasing a couple of grand looking leghorn bantam crosses around pens sufficiently large to create enough area to make the game of catching them as long drawn out as possible.

Luckily there was a lovely stable girl to help, and so we duly rounded up two fearsome-looking ladies, and returned home in triumph, rejoicing in our cleverness at acquiring poultry of such provenance, without a penny exchanging hands.

We were clucking away to each other about what fun it was going to be, how wonderful to have our own eggs, chicken acreage requirements and so on (we have a tiny garden, however, that proved to be more than enough in terms of free range space for organic eggs). Quite mad as I might point out here that neither of us eat eggs, unless they are required in fish cakes or a homemade cake (which never happens as despite loving cooking I do NOT DO bake-off,) so the egg side of things was fairly redundant from the start.

Nevertheless, the following day, we returned to acquire two more of the lovely characters, as I had, in my ill-conceived wisdom, decided that they were a great embellishment to the garden with their beautiful plumage and really sweet bantam like chuckling and chortling and giggling.

We proudly installed our latest mad-cap acquisitions in their very smart hen house which had originally been part of a pair that graced the orchard at The Old Parsonage. Still patting ourselves on the backs, we decided to call them Deborah, Henrietta, Sarah and Hannah. Hannah is the name of the lovely stable girl.

It soon became clear that this experiment was going to go very wrong, for reasons I had not entirely anticipated. They were quite charming, easy on the eye, rubbed along together alright, laid beautifully, but they antagonised Ellie no end, until she actually got to the point she would refuse to go in to the garden, unless pushed. But, horror of all horrors and almost worse, they helped themselves to the plants....

And in what abundance...It is bad enough when Ells helps herself greedily to the lungwort every spring (to the extent I annually trawl Google to find out what possible benefit she could derive from consuming such quantities of this plant only to come to the reluctant conclusion that it possesses the same qualities for dogs as cat-nip for cats, namely it is hallucinogenic).

"Get that dog out of the lungwort!" comes the cry every spring, as hope springs eternal on the part of this little plant but yet again to be defeated by the lurcher's obsessive nibbling. Actually as of writing, we are officially devoid of lungwort in the garden.

This cry was magnified on my part when I began to realise the enormity of my foolish fowl foray as daily I witnessed the onslaught of the slaughter of my beloved plants.

Now, in a very small garden you have very little space and so what space you have is then devoted by default to the very precious few plants you can place there, so one is extremely proprietorial about them. Which does not sit well with four hens,who are quite determined to eat the highest quality diet, despite all the grain and corn you give them, irrespective of the horticultural aesthetics.

Our little garden inevitably became a battle ground, with Deborah leading the charge, as quite the naughtiest, with Sarah never far behind. I sat it out for some time as the feathered additions to our life had their own particular charms. Over the course of that summer, I delighted in watching them run in that peculiar hen-like manner of ladies with long skirts, throwing their little legs out to the sides as they moved, witnessing the interesting social interactions between them, revelled in their generally sweet and low chuckling and for a while there was détente.

Henrietta, in particular, was a social creature who liked nothing better than to creep into the garden room very quietly, when the door was open and surreptitiously glide silently up to my legs, where I would be sitting working at the kitchen table. I would inevitably jump when I felt the unexpected brush of her feathers against my leg.

Now here, I must admit to a curious phobia which I have learnt over time to ignore: I hate handling hens of any type. I am fine with small and even larger birds such as house martins, or sparrows, or even hawks but for some reason I loathe the feeling of chickens. I have learned, for obvious reasons, to ignore this revulsion or fear, as I have generally been the hen herder of the family but it may stem from a time when I was seventeen and trying to earn a penny or two whilst I was still at school.

There was a chicken farm nearby and I was able to cycle there. It belonged to some neighbouring farmers who we had known a long time, and the deal was that I would help in the house, and when necessary pick the eggs from the hen houses for packing and market.

Hen houses is no accurate description for the hen hell I found myself in every Saturday morning (when I did not have school). The air was rank with the smell of death and decay, dusty with feathers, littered with weakened bodies that received no succour from the air and the sun, nor exercise nor movement, nor love nor care nor attention.

This was the first time I had been exposed to animals simply as a commodity; it was to be my last direct contact as such.

I grew up in a farming community where you knew personally the cows you milked and took to pasture, the ewes you lambed, even the piglets who ran through the orchard. The chickens were huge personalities and a necessary assassination of a particularly quarrelsome cockerel was always a sad matter.

But this…this was the first time I came face to face with the horrors of "factory farming". It was horrid, cruel, dirty, nasty, crowded, dusty, smelly and, above all, totally undignified, both for the poor birds, but also I would argue for the farmers. What does it say about your values when you are prepared to sacrifice well-being, both animal and human, for mere profit?

The eggs were plentiful and totally insipid, the yolks a pale imitation of the deep almost bronzed yellow of the healthy bantam, the feathers dusty with neglect and their eyes were sunken in, for truly this was, and all farms like this are, the epitome of hell.

So, the phobia may have its roots in this early job, but found its outlet in giving future fowl the best care and love possible. The experience certainly lies at the root of my somewhat militant attitude towards food sources and the hi-jacking of the food supply, but the only thing we can do is put our wallets and our mouths where our heart is.

The clucking duchesses and Hannah in some way became an expression of that stance, but there was a limit.

They were the most highly destructive birds I had ever come across and for all their sweet natures, they pillaged, plundered and murdered their way through the garden plants, until the day I said to Christopher that we had to call a halt to this Good Life madness, despite the lovely eggs which were proving to be so popular with our friends.

The final straw was my beloved hellebores. The hellebores have been collected over 20 or so years. Despite losing some prized specimens along the way, it is still a reasonable collection, particularly of some doubles which I have grown since they were plugs and which took three years to flower.

Every Christmas I am rewarded by the single pure white Christmas Rose that puts its head well above the parapet long before the rest of the plant. The Christmas Rose, in my long experience, usually flowers at Easter-time more often than not. These hardy plants delight for nearly six months of the year for so little trouble, and yet I think they are some of the most beautiful and varied species in the world. When the chickens decided hellebores were fair game, it was war.

Having been through every annual seedling they could get their little beaks at, including to my fury, the borage, they made dirt baths of every flower bed. After they had woven tunnels and openings through the box hedges so carefully nurtured over the years, attacked the herbs, squabbled with each other, refused to lay in the hen house and generally made the most massive mess and nuisance of themselves, it was time to move on.

Well not quite, it was exile, down the road to some very nice neighbours, who already had a large group of feathered friends. I fear they did not last very long, and the garden was very quiet without them, and perversely, of course we missed them, but it was a salutary lesson that expectations do not always meet aspiration. Hens are only really suited to large spaces, and they are definitely not suited to cramped, confirmed and unhappy quarters. My latest foray was probably to assuage some of the guilt I felt from "picking eggs" for money, but at least this paved the way for a more humane sensibility of our feathered friends. Our charming but naughty feathered friends are much missed but fondly remembered.

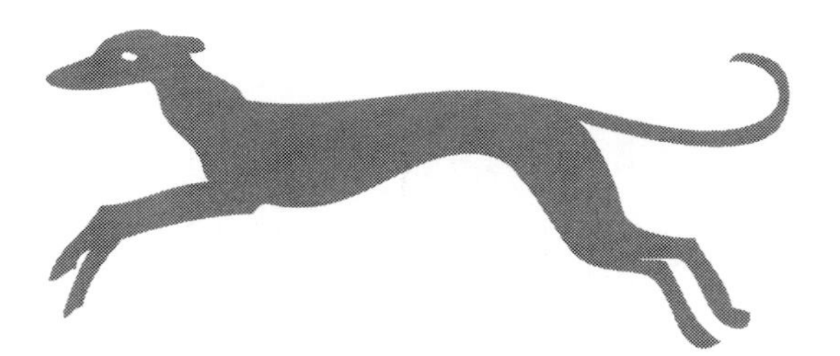

The Garden at Hillesley

Christopher and Pippa

Sunset From The Study

Snow Garden

Frogs In The Irises

Romeo The Robin

Goose-Stepping Pheasants

Highgrove Hen Havoc

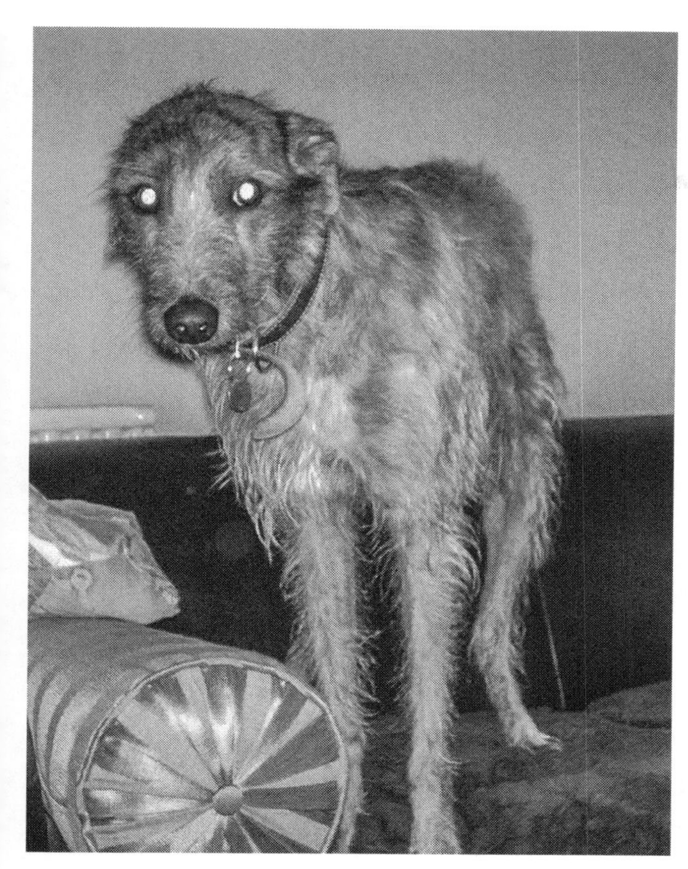

Ellie -
The Results of the Google Search

Image Credit: Lurcher Link

Ellie Trying Out My Father-in-Law's Cavalier King Charles Spaniels' Bed For Size

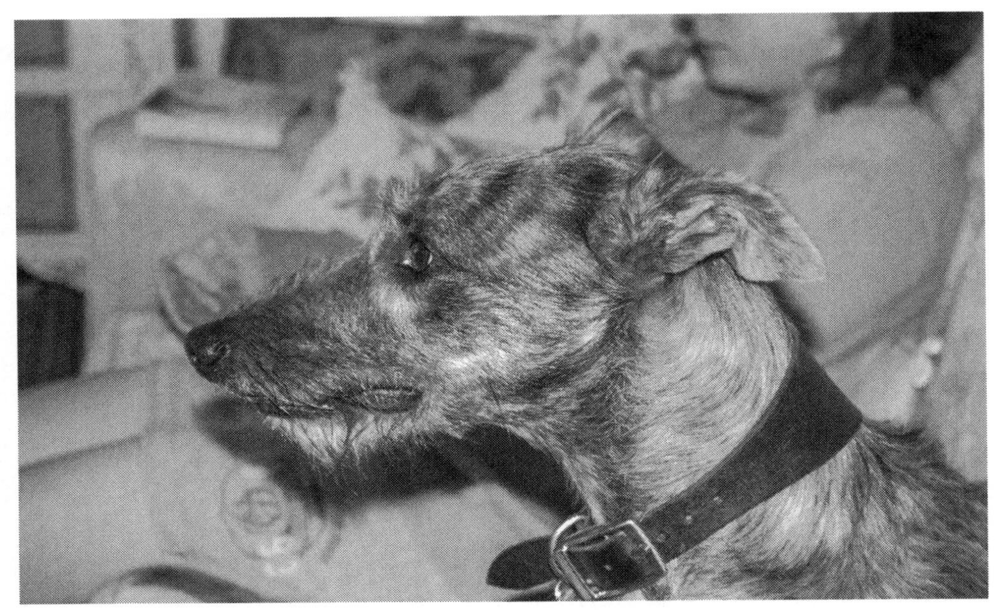

First Night In Gloucestershire - Something Stupid On Television

S N O .....!

I Kissed a Frog

Tapestry Hound

Frog-Hunting

Stone Dog

River Dog

Super Dog

Rhosilli Cliffs

Sandy Bay, Near Aust

Sea-Dog On The Gower Peninsular

Sea-Dog

Lurching and Churching

Pippa and Her Shadow Dog

Intelligence

Foresight

My Most Beloved Dog

But, For Now, There Are Still Adventures To Be Re-told, Relived and Still To Come

ELLIE

CHAPTER 40

Born To Run - Born To Negotiate

Ellie, it became clear very early on, had learnt language from a master. Both her understanding and her spoken utterances surpassed Robbie's, though his speed and agility continued to escape her.

She did, however, manage to unnerve me with regard to her sanity soon after she arrived with us. I had taken her out on my own for the first time and let her off the lead. I generally make a rule that dogs are kept firmly on the lead for the first few weeks until they have acclimatised to the new environment and discipline. Then, unless necessary, they are usually without the lead, as I feel it empowers both dog and owner to share that implicit trust and unspoken line of communication without being tethered.

"Stay close, right at heel, come by," goes the centuries old song of the shepherd and his dogs. Well, Ellie and I went out for our first walk on our own together up the hill in the pouring rain in early September and I let her off the lead.

She seemed very compliant, stayed close by, other than some fast runs which definitely showed her greyhound ancestry, and I very much enjoyed our damp walk along the stubble at the top of the hill behind the house. Normally, you have a fantastic view of the estuary from there, but on this grim day we were lucky to see to the other side of the field, it was so damp and dreary.

But not Ellie's spirits, she trotted along, ran joyously and came back to heel nicely and we went home gratefully. I was in a jolly mood, as I was delighted with her behaviour, that is, until we got into the house.

I had not bargained for the heady effect of relative freedom upon the newcomer and, as soon as she got into the garden room, wet and muddy and generally dishevelled, she took off round the room like a little rocket. Moreover, not apparently around the floor but the WALLS, having got up speed by jumping on the sofa, disgustingly dirty and sodden, she proceeded to circle the room by bouncing horizontally to them.

Mud sprayed everywhere, but I was helplessly rooted to the middle of the floor, frantically worrying that somehow we had acquired an entirely mad dog who was completely out of control. The only comparison I can possibly draw to this near impossible feat of muddy athleticism is the motorcycle "Wall of Death" where velocity keeps the rider safely glued to the wall.

As suddenly as she had started, she stopped, and flopped onto the sofa. Considerably less muddy and damp, as most of this had been dispersed around the walls, so they looked as if someone had artfully applied a thin layer of fine brown mud and thoughtful splattering created by the speeding vortex. I dried her down, but somewhat dampened in mood, as I was considering whether we might have acquired the loopiest dog in England.

She never repeated this performance inside but outside she has for many years (at her own whim and never on command) executed perfect 360 degree turns and heady air borne loops. Even now she delights racing through the muddiest dirtiest puddles at the highest possible speeds. Oddly though, she always insists on being towelled down or even washed on arrival home as she is almost fanatically fussy about her cleanliness in the house.

I can only assume that day that she had finally realised she was here to stay and was expressing, in her own, way happiness and excitement at having a proper home of her own again; well one thing's for sure, she painted the walls brown!

CHAPTER 41

One Wedding, Two Lurchers And One King Charles

(Spaniel That Is)

Or, The Day The Snow Fell

December 18th 2010 4am.

The long awaited day – our wedding.

"It's snowing."

"What??"

I rushed downstairs and opened the garden room door to see an enchanted land; everything was covered in snow, deep, inches deep, white snow.

And it was STILL snowing.

I love snow, I mean, I really, really, love snow, as I have described with Robbie. However there are circumstances in which I do NOT like snow.

A light aesthetic dusting, maybe, a Christmas card icing, a Swan Lake frosting, perhaps, for the photographs. But not this full on, gently twirling, twisting, furling, unfurling and relentless emptying of Mother Goose's pillows from the sky; not today, please, not today of all days.

C went back to bed (men are always so helpful in these circumstances) and I sat, and I sat, and I sat, and watched the snow, falling so softly, so silently, so beautifully and so unremittingly.

"Snow was falling snow on snow,

Snow on snow…"

I began to wish with all my heart that I could stop the snow.

All my life I had wanted a white wedding, and the last carol in the order of service was *In the Bleak Mid-Winter*, by Christina Rossetti and quoted above. You have to be careful what you wish for in life, don't you? I ruminated glumly to myself.

It was becoming clearer by the snowflake that a 4pm Wedding was no longer on the cards, or possibly any wedding at all, if the snow was to continue at this rate. All the planning, all the anticipation, all the food, even the lack of a proper wedding dress, and the few people who actually had been invited, were slipping away from this bride, silent flake by silent flake.

At about 7:30am, the snow started to thin out, and then come slowly to that graceful halt, so reminiscent of the final bow before the curtain comes down at the ballet and that was it. Light began to appear and reveal the magical sugarplum world, which clearly would be quite un-navigable for the majority of our invited guests.

It was just eight o'clock in the morning when I phoned the Rector.

Now let me introduce you to one of my most favourite people, other than my husband and my girl, that is.

CHAPTER 42

THE RECTOR

When I was a little girl my favourite book was *A Little White Horse* by Elizabeth Goudge. There were so many things I could relate to in it or dream of: the lion dog, the marmalade cat, the salmon pink geraniums, and the estate of Moonacre, which stretched from the moors to the sea, the dark wood and...the old parson, who actually turns out to be French and rather dashing.

Well, this parson completely flummoxed our heroine, and perhaps, I am not quite sure which heroine we are talking about here, as fiction once again begins to blend with fact. In the above mentioned book, the parson is exactly as I was delighted to find here in later years, a real parson, charismatic, clever and charming. Maria, the heroine, had been so entranced by this officiant that she had imagined, and it felt to her as if, the roof were lifting from the church at his command. The services were characterised by the sonorous boom of his voice and the searing wit. Which all made for the most delightful Sunday morning, and the partaking of church, no chore but pleasure, as if all the angels were singing in heaven too.

Our Rector, who as you will gather is of a similar ilk, was not yet awake that morning when I rang to re-arrange the service due to the weather, but his lovely wife answered and soon afterwards the wedding was moved to midday.

Even that wasn't enough to save the day, as the snow was moving eastwards, and just as my father-in-law to be was preparing to leave Kent the snow hit, and meteorological circumstances dictated he abandon his plans to join us.

The Rector had had the foresight to put together a recording of the music that we had in the order of service, as the snow had been predicted (but not perhaps its severity), and the organist was unable to reach us. At least we had some music.

Well, I went upstairs to raise my snoring husband to be, and my slumbering daughter, to tell them of the change of plan. They hastily arose at this point and I hurriedly put together my white rose bouquet, and C's buttonhole. The food was hurriedly loaded into the cars as no one would be able to reach the appointed reception venue and at 11:30 we left the house and gingerly made our way the scarce few miles to the church.

Needless to say, by this point, the sun was shining brightly and the snow glistened and Badminton Park looked for all the world like an enraptured land.

However, we knew we were down to 19 guests out of 50 as the morning had been spent calling the guests to tell them of the change of plan and many were snowed in, like my father-in-law. Some of my family made it, albeit in wellies, and a few locals, but we were thin on the ground so we moved to Plan D.

CHAPTER 43

PLAN D

Plan D was to prove more entertaining than had been anticipated and would lead to some of the many developments outlined in this little book.

In my earlier telephone with the Rector, I had asked whether, given the paucity of our guest numbers, it would be possible for the family dogs to come to the wedding in the little church, to which he readily agreed.

So, in order of appearance, came Ellie to whom you need no introduction, Amor, my mother's lurcher and Georgie, my brother's King Charles spaniel. All semblance of formality having been derailed by the weather, we all trouped into the church together; a motley crew, it's true. But more was to come.

The opening hymn was *Guide Me, O Thou Great Redeemer*, and then the roof of the church seemed to open to the heavens but not with angel voices, but rather the most unexpected choir you have ever heard. Just after the opening line, the dogs joined in with our singing, Amie leading the way as tenor, Ellie maybe not quite so proficient, and Georgie, a delightful soprano.

All sense of decorum at that point was lost, the frustration of the weather dissolved and as the Rector's wife said to me immediately after the service "I think that that was the nicest wedding I have ever been to".

It got better: the wedding breakfast, was held not in the church as anticipated, but at a friend's house, the dogs continued to enjoy themselves immensely and great fun was had by all.

As you will read next, the 18th of December often is rather a dramatic day for us, but so far has not surpassed the day the snow fell and the dogs sang and the roof of the church was lifted to Heaven, doggie Heaven that is! I have often wondered whether I was wise to insist on *In the Bleak Midwinter* as the final carol, for snow most certainly had fallen, snow on snow, that day.

CHAPTER 44

DEER AND DRAMAS

Our first wedding anniversary and I was preparing a dinner party to celebrate: I looked at the clock at five minutes to three: food check; table check; flowers check; checklist check, and thought "Great, everything is in order for dinner tonight and I can go and have a little rest…"

I did not completely finish this thought, as another more urgent flashed across my mind. I rushed to get my boots and Barbour, grabbed the phone as it rang.

"I think you need to come and look at Ellie," said the voice of my husband.

"I am already by the front door, already," I said.

The car pulled up at the front of the house on the road and I slammed the front door behind me and ran to get into the car.

"What happened?"

We still don't know exactly.

"I don't know, she ran off and came back like this about 10 minutes later. I had heard an almighty squeal and seconds later she returned to me."

"I hope Rowe's (our local vet) is open."

Four minutes later we were getting worried; Rowe's was not open. it was the Saturday before Christmas. I started to panic,

"Dursley, quickly!"

Dursley is some 11 miles away and they seemed the some of the longest of our lives. I had hardly dared look at Ellie. I am used to seeing pretty awful accidents, breaks or even picking up run over cats of ours, but this was horrific, worse than anything Robbie had done.

Her left rear leg was slashed about three inches across the knee joint and splitting the skin. You could actually see the bone and the tendon. Oddly enough there was little bleeding, but I could see the pain, the fear and the terror in Ellie's eyes. She couldn't even sit down and I found it almost too difficult to inspect the wound. Her pain was our pain.

Luckily there was an emergency nurse at Dursley, Vale Vets, and she took one look at Ellie and paged the vet on call. Ellie was on the operating table within twenty minutes and received the most fantastic care.

The dinner party was a great success, but it passed in an oddly dreamlike way for me, and I suspect C, as half our attention was with our girl, willing her to get well again. The injury was so extensive and deep that I feared that she might be lame. My fears were groundless, by some miracle the blood supply had not been punctured, the muscle and tendons were intact, though the gash millimetres away from them, and the bone unharmed.

She lived to run again, faster than ever, but Christmas that year was a distinctly muted affair.

Not content with that drama around Christmas, last year in 2017, Ellie developed very severe Pancreatitis. To watch her in agony, and not be able to help, was one of the worst experiences of my life. Oddly, it seemed even worse than the leg, because we knew that once she was with the vet, she was likely to recover, as long as there was no infection. She was also a young and very fit dog.

This time felt different, she was ten only days previously, she was clearly in agony and very frightened and we did not know if she would pull through. After three nights in hospital, she returned home. We were all much chastened by the experience, and now we are even more watchful of her diet.

Christmas, though, was truly a time of thanksgiving for our beloved girl, who for now, was back to her usual bossy, and inimitable form.

CHAPTER 45

Little Donkey

Ah, the nativity play, one of the high lights of any proud parents year, though at a couple of M's early schools they seemed rather keen on an "alternative version", which totally deprived me of the immense parental pride of seeing my little girl looking angelic if only for the time of the performance. Not that I wasn't proud of her in her plays anyway, and she has done lots and I have enjoyed them immensely, but oh how I longed for the traditional Nativity Play!

I particularly enjoyed one in a church in Cirencester which was held for the residents of an Old People's home and the local Special Needs School. I had the privilege of helping the old people, and it was such fun to see all the faces lighting up in the church when we witnessed the miracle of real sheep, goats (you have to be kidding), chickens, rabbits, and other assorted animals and a beautifully attired cast from the school with some very angelic angels and regal wise men. Best was the entrance of Joseph with Mary on a REAL Donkey!

The performance was made somewhat more interesting for me, as my mother had also brought along her lurcher, Amor. They were seated right by the sheep and Amie was exceedingly disconcerted by this proximity, but it proved only to be a Mexican stand off and everyone retired home unharmed!

My overwhelming maternal pride at a Nativity performance came quite unexpectedly though. Our local church here, in Hillesley, was giving its annual Carol Service which for some reason I decided to attend on a whim. I asked the Church Warden whether it would be alright to take Ellie into the church, despite the notice on the gate which says "No Dogs in the Church Yard", and he graciously conceded.

Some of the children in the church were fascinated and somewhat distracted to see Ellie, but she had an unanticipated opportunity to show her worth even more. Ellie, as you may have gathered, is quite a practiced church-goer, like Amie, so I was totally unconcerned as to her behaviour despite the children's curiosity. What happened next took me totally by surprise.

When the children had duly lined up at the altar to give their little Nativity Performance, the Vicar said:

"Oh, the only thing we seem to be missing is the little donkey."

I do remember other animals being present in the church but none of them appeared to have the colour or size that might approximate a very small donkey. I said before I could possibly stop myself, being on occasion a somewhat impulsive person and recalling immediately how when Ells is sulking and not getting her own way, she bears a remarkable resemblance to Eeyore.

"Actually we have!"

Ellie was duly invited to approach the Chancel and I took her up on her lead. What happened next astonished not only me but I believe many present. Ells lay down by the altar without demur and proceeded to stay utterly still throughout the performance, even though I had retreated to the aisle. She did not do any of her normal "I belong to the local rent a dog group" nor hound the children in anyway.

Clearly a born performer, she stayed quietly right in front of the altar for the whole performance, and at the end retired most gracefully back to our pew. Being a brindle grey and quite large, it would seem that the children found her a most satisfactory substitute for a real donkey.

To the extent that one little boy, who was getting fractious as it was way past his bedtime so his mother decided to take him home, said very clearly on their exit:

"Good bye, little donkey, good night."

Waving all the while, the child departed in his mother's arms. My parental pride in my nativity star performer knew no bounds. After all those years I had finally seen my little girl, albeit the canine one, star as brightly as any angel in the school nativity play. It truly was a great Christmas present for me and I was told that everyone else enjoyed her involvement too.

CHAPTER 46

Lurching And Churching, Meeting And Greeting

I think it is possible that in Ellie's mind the church is indivisible from the pub as a meeting place, though to be fair she does seem to recognise the need to be quiet in church, unlike in a pub, where she will talk to everyone and anybody.

We take her to church more often than the pub. (Uncle George takes her to the pub on any possible occasion! And she loves it.) Only say the word "PUB" and the little meerkat ears go up like veritable antennae, seeking out the nearest possible location, to the extent that she even recognises some of the local pubs when we drive past and the look of disappointment on her face, when we do not stop, is hilarious.

Even worse than not stopping at the pub though, is leaving her behind when we go to church which is quite often as my husband is churchwarden to two local churches: Great Badminton and Little Badminton.

The appeal of each is probably different, we gather: Great Badminton was for a long time a place shrouded in secrecy for her as it is approached down a long drive guarded by big gates which mysteriously unlock at the tap of a few fingers. You cannot even see the church until you have walked along the path bounded by hornbeams on one side and a ha-ha the other, looking across at the most glorious pastoral view of horses grazing amongst the picket fencing. It is a scene utterly reminiscent of Stubbs' painting of the English landscape.

Oh, the excitement when she was first allowed into this enormous building, especially the length of the aisle, and the odd wooden pews and smells, the musty vestry, the cold and sometimes damp smells. But the mystery there is that there are never any people, at least that must be what she thinks, seeing no-one but us.

Little Badminton is quite different:

There is a large graveyard to play in and snuffle through the grass and potter along the side of the church as if for all the world it is her own garden. It reminds of a story of a vicar who was being visited by someone in his parish who was found in his graveyard, as it were, and hiding behind a gravestone, looking very shifty.

"What are you doing, Vicar?" said the parishioner in an astonished and somewhat disapproving tone.

"Why playing hide and seek with the dog of course," replied the Vicar insouciantly with an air of complete innocence, for he was indeed being entirely truthful, apparently much to the chagrin of said enquirer who was probably taken most aback.

There is another charming anecdote about a vicar, who asked how he was lucky enough to grow such wonderful roses, replied that the cats were coming up roses. On further enquiry, the questioner was enlightened by the further explanation that the former feline residents of the Vicarage, whose job it was to keep down the church mice, were buried under the rose bushes.

I digress, the best thing about Little Badminton is that Ellie likes to meet and greet people at the door. She is a most sociable and charming animal where people are concerned, but less so around other dogs, unless of course they are either family or friends or long dogs. The hierarchy of the canine social distinction and scale never ceases to amaze me. The snobbery and air of being set apart is not only precise, but has also been a characteristic of all of the lurchers we have enjoyed.

Herein lies the similarity of the church to the pub – there are people to talk to, though the pub comes with the added possibility of a morsel or two as well.

So imagine the times when we are getting ready for church and we have to tell her "No, not this time". And hear as the protracted sigh, that proclaims the beginning of a sulk, is slowly exhaled as her head slumps down on her paws on the sofa.

She has had some reasonable excitements and outings in her time to church, and as you know she, with the other dogs joined in our wedding celebrations with great enthusiasm. It is lovely that around here,with few exceptions people are delighted to see her, and take the view that dogs are all God's creatures too. We actually had a dog that came to church regularly for the service, called Bella, an elderly but charming little terrier who was a great friend of Ellie's and there was Badger, another lurcher who she meets in the church yard, and Pepsi, the spaniel, and so on. In fact it is a charming sight to see the meeting and greeting and exchange of news with them all.

When I was a child, I was peremptorily informed by a Roman Catholic priest, in our drawing room at the Old Parsonage, that animals do not under any circumstances go to Heaven. I am not afraid to say that I renounced my intention to go to Heaven, much to the horror of the assembled adults. I have always adhered to that opinion, heaven without animals, particularly dogs, would be hell to me. What was the point of the Ark, then? I am sure Noah would agree with me.

One day when the first of us has shuffled off our mortal coil and is laid to rest in our churchyard, I hope, then Ellie will be there too. We can never think of a more pleasant and peaceful place to forever play hide and seek amongst the gravestones and the old yews and the spring flowers. One hopes and prays for the continuity of this way of country life and parishes, of welcoming worshippers and forgiving vicars, as we are all God's children in my book.

CHAPTER 47

Bed And Morning Cuddles

Ellie mostly sleeps with me in the autumn and winter, though she is a famous bed-hopper and can ricochet between the two of us throughout the night, leading to somewhat disturbed slumber for all concerned. She is less keen on my early morning rising, as I rarely sleep past four or five o'clock, particularly in the Spring and Summer when I am woken at first light every morning by that raucous young blackbird determined to get in first with his early morning song of praise to the new day.

There are times where it would appear that his family do not appreciate this early alarm call, for I often hear some distinct grumbling from his elders, and get the distinct impression that they are telling him to "Shut UP". Then they put their heads under their wings until some three quarters of an hour later, when the garden becomes a symphony of sound. It begins with just that single pure note and swells into a crescendo that defies even the most determined snoozing…First, the blackbirds with their choir like pure tones, then cheeping sparrows, soon after the swifts' sharp cries as they race through the deserted dawn streets, the pigeons and ring collared doves with their cooing and billing. Then bubbling sound of the house martins and the swallows is added to the mix, the alarm cry of the pheasant in the wood, the distant crowing of the cockerel and, on the hill, the final trill of the skylark ascending.

Ellie's morning mutterings are somewhat more prosaic, along the lines of "Do we have to get up? I am comfortable lying here, let sleeping dogs lie", and so on…until once more she is enticed to rise by the promise of a walk and running again under the open sky…

She is, though, very partial to a proper early morning cuddle and, even as I am writing, as I mentioned at the beginning of this little book, a paw will snake out and hit my laptop…go back to sleep or stroke me…I kiss her good morning and hold her close and breathe in deeply that most glorious of scents, a healthy happy dog. Until she suddenly remembers Christopher and jumps off to go and remind him that it is a new day and he needs to get up, which is not always so successful, as my husband prefers to sleep more civilised hours. So she goes back to snuffling and dreaming and muttering those little cries of the hunting reveries which punctuate the canine slumber.

CHAPTER 48

FOOD

All my lurchers have exhibited an idiosyncratic approach to food. Robbie and Benjie and Ellie, in particular, have tended to a) be very picky about what they would eat; b) dictate when and where they would eat; c) show great disdain at any offering which they deemed beneath their attention (biscuits and cake as I described earlier with Robbie); d) perk up at the sound of plastic wrappers which might contain some thoroughly unsuitable and addictive doggy treat or even newspaper if it contained fish and chips; e) know exactly who to ask for what: Mummy for proper food, Daddy for chips, Uncle George for Hammy Whammy, Grannie for anything you could get (except biscuits!).

None of them have been thieves, though a relation of Benjie's, who lived not so far from here, was famous for eating a luncheon prepared for Royalty. The lunch had been laid out on tables in the garden, perhaps optimistically, but imagine the embarrassment when the luncheon party arrived to partake of their repast to find that someone had been there before them. Moreover, that someone had done a jolly good job of clearing up everything but the salads. This culinary disaster was taken in good humour by the regal guest, who apparently just rolled up her sleeves and made scrambled eggs for the assembled guests.

One often hears of lurchers who spend the night in the nick and have more than a passing acquaintance with the local police, due to their habit of appropriating food from any available source. Sadly, this penchant for theft has not been unknown to lead to an unhappy demise.

I know of another canny canine, who lived in a very grand stately home, where there was an Italian Restaurant opposite the end of the not inconsiderably long drive. Foreswearing the snack hut by the lake, this cunning animal would sneak out every morning at opening time. and spend his days at the restaurant, receiving scraps gratefully until returning home late afternoon. It would then partake of a jolly good dinner, before retiring for the evening. Said dog grew very plump indeed, but nothing would deter him from his daily excursions.

Ellie, though, is a great deal more fastidious, certainly about her dinner. Once she has reminded us that it is very important that it is made at the appointed hour, she retires to bed for a pre-prandial snooze, coming down when she chooses to cautiously inspect our culinary offering for that evening.

Approaching the said bowl, as if convinced it must be poison, she usually retreats to consider whether it is worthy of her attention immediately, or whether it can wait awhile until we are eating too. I might add here that this is no ordinary dog's dinner of a tin and some kibble, but usually roast chicken and vegetables specially cooked for her with homemade stock, or if she is lucky some beef or lamb or left over game depending on what we might have had.

When I once mentioned to someone that I cooked the dog's dinner daily, they looked at me quite astonished and told me I had created a rod for my own back. I retreated hastily from the conversation, somewhat perplexed for this person adored her dogs, and I likewise, which is why they do not have tinned food and biscuits.

Whilst this was not the case when I was young, and the animals were fed more conventionally, I became aware of the dangers of commercial animal food when Robbie fell very ill when we living in Bristol.

I came back one day from my shop and opened the door to the most appalling stench and rushed into the kitchen, to find Robs lying supine on the floor in a pool of watery sick, and worse. Realising this was a total emergency, I picked him up as he could barely walk and carried him to the car. I rushed to school where I was supposed to be collecting M and asked if they could keep her late as I had to find a vet urgently. They kindly looked after her until I could collect her later.

Panicking somewhat as Robbie's normal vet was fifteen miles away, I dashed to the nearest available vet. I was telling him to hang on in there, as his condition was deteriorating very quickly, he was clearly in pain and horribly dehydrated. I couldn't believe this had happened so suddenly in just the five hours since I had left him!

They were wonderful at the Vet's, calling the senior partner down immediately. She happened to be a famous TV vet, who has I believe returned to her native land, but that didn't matter at that moment. She took one look at Robs in the car and we carried him into the surgery together.

That night he had two litres of IV saline and antibiotics, and obviously urgent bloods were taken. An extremely expensive exercise but luckily I was insured. He turned out to have a gastrointestinal problem, so once he was reasonably recovered, thanks to their marvellous care I took him to see Nick Thomson, a highly accomplished equine vet who also practices animal acupuncture and homeopathy in Bath.

He advised me not to feed Robbie standard dog food, but to put him on the BARF diet, which is basically a raw diet which includes pureed vegetables and raw bones. Well, I had already improved his diet after the emergency by giving him chicken and rice to help him heal but he never took to raw food or bones at all.

Nick also treated Robs with acupuncture and Pulsitilla, the homeopathic remedy, for his increasing stiffness of gait and weakening back legs, and his growing neediness. Robbie was then nine, he went on as you already know, to a very good age of fourteen and a half before leaving us. A true testament to a good wholesome diet.

So back to the rod for my back, if you examine the origins of the tinned commercial dog food market you will find that this was dreamt up by the commercial merchants of the so called "Green Revolution" in the 1950s after the war. This was when the world's food chain started to be commercialised heavily, and the business owners were looking for ways to decrease losses from farmed meat, and increase profit. This has led to the increasing obesity rates and chronic ill-health in animals and men that the world is currently experiencing.

They came up with the brilliant idea of hamburgers from offcuts, and with whatever was left over and even more scrappy. Tins for pets. If you examine the ingredients of any commercial pet food, you will notice that invariably it will say something along the lines of 30% or 60% beef or salmon or duck or whatever delightful flavour they have devised. The rest is fillers. This is now mixed with grain-based kibble! However, our feline and canine companions evolved to be carnivorous with the occasional foray into ripe fruit or vegetables; wolves for instance eat goji berries, foxes eat grapes, some of our dogs ate raspberries and Ellie likes dates, like her desert counterparts, the Salukis.

Clearly this new-fangled nutrition did not nourish Robbie, and I suspect not many other animals, given the rates of disease in our domestic animals. Ever since then I have tweaked my animals' diet and given them home cooked food with occasional treats. Ellie regularly eats raw bones, provided by our wonderful local farm shop, and her teeth, once terrible from being fed slops, are now excellent, white, clean and strong. She does not care for raw meat unless it is very fresh mince, so I give her roast chicken or cooked mince or sardines with some sneaky vegetables.

So it is probably the vegetables which she is checking out suspiciously, but sooner or later, she succumbs to hunger and polishes off the contents of her bowl. It is another of her idiosyncrasies to leave something for later or, worse, if she does not approve of my offering, she leaves it with an air of disdain, and there the food may languish until the next morning whereupon it may, or may not, be consumed.

This appears to me to be closer to the eating habits of animals in the wild, who certainly do not eat when they are ill, and I generally have fed my animals on demand, as I trust them not to overeat and none of them have been greedy. (Except for cheese, chocolate, sausages and ham!)

The rod for my back, it seems to me is simply the same nutritional care we give to our children and our spouse, so why not our pets, who are surely part of the family? To say nothing of the vet's bills saved, the illnesses avoided and the joy of witnessing joyful and healthy companions living their allotted span and sharing with all the happiness they enjoy.

CHAPTER 49

SPEECH AND SYNTAX

Badminton, Badminton", she was saying quite clearly to the astonishment of my friend and both of us. "Badminton", she reiterated, staring at the coverage of the cross-country phase of the famous horse trials being shown on the television.

Earlier that day we had come back from visiting the Horse Trials where we had taken only our friend's dogs, leaving Ellie behind to total doggy fury and pointed sulking. After lunch she was quite so clear about the unfairness of the morning's excursion, even to the point that she seemed to be stamping her paw, looking at the television and repeating "Badminton" in an increasingly exasperated tone, that Christopher and I gave in and went back up to the cross-country.

Generally, I prefer to watch the cross-country on the television, as one can see so much more, but Ellie wasn't having any of it; she wanted to go and see her beloved Park, where she had witnessed the slow build-up of the event for months.

For ten months of the year the Park is very quiet, a few dog walkers, some anglers, Estate traffic and the occasional single day event, but from March onwards a veritable village is built. Each year it is a source of surprise to Ells that there should suddenly be so many people, and dogs, and horses, to say nothing of the cars, the crowds, the stands and the aroma of food, everywhere.

One of her favourite treats is to go and visit one of the best trade stands, which is run by a cousin of my husband. Always assembled in impeccable taste, in the style of a country house interior, there is always a really good atmosphere on the stand, as if there is a party going on - and Ellie loves nothing more than a party.

She always makes herself very at home, saying hello to all the girls who work there, to say nothing of her greeting to Charlie, of whom she is very fond. Somehow, it suits her, the party atmosphere, against the backdrop of leather boots, fashionable tweeds and the clinking of the glasses at the bar.

She does not seem to be quite so keen on watching the event though, sniffing somewhat disapprovingly at the size of the jumps, and a little frightened, maybe, by the sheer speed and majesty of the competing animals.

What is noticeable though is the large number of long dogs at these events: sometimes it would seem there are only a very few breeds in the world: black Labradors, Spaniels, Deerhounds, Wolfhounds, Terriers and Long Dogs of many kinds make up the vast majority of the attendees!

Coming back to canine communication, she also has a very practiced line in telling us in no uncertain manner that we have to go to "Wotton". She is partial to a treat, some disgusting-looking fake bones purported to be good for her teeth, but which must be the doggy equivalent of McDonald's for their addictive qualities.

She demonstrated this ability quite recently, when we had made it abundantly clear that there was no trip to Wotton that evening for said treats. Subsequently we were treated (if you will excuse the pun) to a most eloquent and seemingly syntactical speech. This entailed both, what were quite evidently long sentences of explanation, as to why the trip was absolutely necessary, and a great deal of graceful bowing down of the head, with a back arched upward in a bow shape with a purposeful wagging of the tail and this sentence is so long because that is how she was talking.

She won – the display was just too funny; she knew it too and, somewhere in that clever mind, she will have stored up the performance for another, necessary, occasion for a visit to "Wotton".

Robbie was not quite so vociferous, though equally good at getting his point across, not only collecting his bowl or lead, which is normal in my book, but worse, being a back seat driver.

One time I was visiting friends who lived near Aust, where my old home was, and Robbie and I were going round that slightly famous roundabout just on the English side of the Old Severn Bridge, when I heard him give the most indignant snort from the back seat. I gathered pretty quickly that I was being informed that I was very clearly going the wrong way!

You see, when you approached The Old Parsonage that way, you would take the 3 o'clock exit off the roundabout, but I had committed the cardinal sin of going straight on over. The backseat disapproval was palpable, that is until on arrival at our destination, Robs recognised immediately a spaniel there to whom he taken an instant fancy to some months before and cheered up instantly.

Benjie also was a great communicator, partial to huge and prolonged sulks whenever I told him I had to return to Cambridge or London, or wherever. He was the first master of semaphore too. The most memorable occasion was in the 1990s when my parents, my fiancé and I had taken him with us to walk in the Black Mountains in Wales. Having had a long and extremely muddy walk we returned to the car where Benjie proceeded to tell R, in no uncertain terms, with a peremptory and commanding nod, that he expected him to get in the boot of the car and that he, Benjie, would reside on the back seat with me for the journey home. This did not go down well at all with R, who was at that time used to much smaller dogs, though still of great character, but I am sorry to say my parents and I found it hilarious. Benjie proceeded to insert himself very firmly between us on the back seat and we endured an extremely cramped and uncomfortable journey home.

CHAPTER 50

Semaphore And Sleep

Ellie, too, seems to have mastered the art of semaphore, though it sometimes is something of a guessing game as to what exactly she is trying to say. Eyes to the side mean, "Outside", a dismissive flick upwards means, "Don't be so stupid", a sudden slow blink "Maybe", and so on. The idea is you ask her the question, and she signals her answer, as it rare for her to be quite as eloquent as I have just described; this is saved for those occasions when we are being so obtuse in our understanding, and the matter so urgent, that extra effort is required to get her point across.

As for sleeping positions there is actually nothing unusual there, if that is you are already acquainted with idiosyncratic body contortions so beloved of the average lurcher. Their long backs, coupled with equally long legs, apparently require the exercise of doggie yoga to get comfortable.

Beds, not baskets, at bed time; if confined to basket at someone else's house, you curl up in a tight ball to demonstrate your sadness and offence at your poor treatment and refuse to uncurl under any circumstances, oozing "Alas poor me – all alone in my basket", from every pore. This behaviour unfortunately reaps no reward in terms of more comfortable sleeping arrangements, unless you are at home.

Likewise chairs and sofas, at somebody else's house, you stay on the floor.

At home you lie upside down on the sofa, legs pretty much straight up to the ceiling, head lolling in a most alarming manner over the edge, and eyes occasionally flickering to whatever may be of interest on the television.

"And stretch", comes the sound, as the elegant limbs extend to their fullest extent, with a deep sigh until there is a sudden exhale, and most gymnastic roll from the back position, and if you are lucky, you land safely on all four paws on the floor. Although, it has been known that due to a serious and highly embarrassing miscalculation, you somehow roll off the sofa in an undignified way, and walk off looking astonished at your own ineptitude.

(And that's the dog.)

Truly, the lurcher adopts the most extraordinary arrangements of their limbs in a bid to get comfortable, and comfort is definitely one of life's primary concerns. Any bed will do but the best is king-size preferable with newly washed sheets, recently warmed by your human hot water bottle, and pillows are a total necessity. Rare is the photograph in the glossy magazines or the Sunday papers where the lurcher is anywhere but lazing on a grand sofa, a fabulous bed or outside by a picture-perfect outside lunch table groaning with sumptuous treats.

This is a life of luxury, interspersed with muddy walks and extreme speed. Unless you get to the lurcher immediately on entry, the mud is well dispersed throughout the house.

To be fair to Ellie, however, she waits expectantly for her towel to be brought for her damp or wet feet and head to be lovingly rubbed down, BEFORE retiring to aforesaid sofa or bed and the long sigh of "I'M bored" or "I'm tired" ensues prior to a little snooze before enquiring where the next source of entertainment is likely to be. I am ashamed to say that we have been known to resort to that modern babysitter, the television, on a wet and windy winter's afternoon with a DVD from our extensive animal film collection.

And so to rain, that bane of any dog's owner's life.

I have never yet encountered a dog who is not absolutely, totally and, firmly convinced that you, and only you, are responsible for the rain and that that responsibility confers upon you the ability to turn off the rain, like you turn off the taps.

In other words you are GOD - accountable for the weather, the food and the comfort of the canine environment.

For which they reward you with love, entertainment and sometimes sympathy, if that does not clash with their own agenda. Truly, I wonder why society talks about the domestication of animals when I think it is very clear that it is dogfication that has revolutionised humankind.

CHAPTER 51

DOGGIE PECULIARITIES

(And Catty Curiosities)

One likes biscuits, chocolate, cake, another loathes them, none of them steal, all of them have been highly idiosyncratic, but let me describe just a few of the eccentricities exhibited by these characters, over the 35 or so years, I have known lurchers.

ALL of them like their hair being brushed in the morning when we brush ours, even better is the full works, in terms of being prepared to go out visiting, or, even more exciting: the arrival of that most favourite of pastimes, guests.

Prior to meeting and greeting, one must have a face wash, Robbie and Ellie in particular. There is no privacy in my bath room here now, as soon after I get into the bath, there is a determined shove at the bathroom door. Whereupon, I am presented with dinner stained, persistent, grey snout to be attended to, and then dried off with a matching grey towel. Whereupon, Ellie disappears for her post-prandial nap, on whichever bed has taken her fancy on this particular evening. Woe betide us in the wet or cold weather should we forget to do the feet and head and if necessary the back with the towel and a special back rub.

But this behaviour has not, in my experience, confined itself to the dogs, as our cats learnt pretty quickly to hop into the bath when it had just finished emptying, as then it is still perfectly warm and damp. Where better to have a really good wash? Indeed, my mother had one cat who was known to have sat on the edge of the bath when she was bathing, and who, on more than one occasion, overbalanced and fell in, much to its embarrassment.

Going back to guests, in the days when women still wore fur coats, we children would be charged with taking people's coats to the spare room after they arrived. On their departure, we were required to retrieve the correct coat for each person at the end of the party. This became something of a game, as my parents liked to give large parties, and all the coats would be piled up on the spare room bed. Amongst the piles of coats, which could look like the scattered wardrobe of Cruella de Ville, as the correct coat was being retrieved, I would be unsurprised to find a warm fur coat of the sleeping variety. The cats liked nothing better than to sleep in the middle of a pile of luxurious furs availing themselves of the comfort. Cruella would have had a hey-day, but I was tactful enough never to inform our guests of the late tenants of their coveted garments, although sometimes one would comment, "Oh, isn't lovely and warm!"

Guests – the perennial excitement of watching the preparations, the extra food eagerly eyed up, probably in anticipation of extra rations later. One of our dogs, a spaniel, would do the rounds at drink's parties, hovering the floor industriously to catch the crumbs from the canapes. Those evenings were marked by the laying of the table, the early dinner of dogs, the lighting of the fires, and doggie anticipation of some extra treats.

Waiting breathlessly for that very first and so exciting KNOCK at the door, the dogs would rush to greet whoever it was. Sometimes disappointing if the said arrival did not like dogs, but generally very satisfying, in that entry to house was granted on the basis of being petted and generally admired. Benjie was funny though, he loved, like the others to meet and greet, then like all the others sleep under the dining room table (they are not allowed to move until the repast has completed). Until, that is, he decided it was time for people to go home. He had a number of strategies for this. One of them was to tug my mother's hand and try to remove her from the table, but one of the most successful was to retrieve, from upstairs in the laundry pile, some incriminating item of clothing which would cause great embarrassment amongst the ladies present and highly effectively remind them of the time. Magic! They would vanish pretty quickly soon after.

This cussed behaviour is actually quite characteristic of these highly intelligent group of dogs for, as we have discovered, they consider themselves a breed apart. For instance, if you say to a lurcher, "Go and sit in your basket", they will invariably go and lie on the sofa. If you use reverse psychology and ask them to go and lie on the sofa, they will, almost as invariably, take it upon themselves to retire to their basket.

Ask Robbie to "Sit Down" in the back of the car, he would ignore you; ask "Robbie, do you have a problem with sitting down?" Then, he would promptly sit down, as if butter wouldn't melt in his mouth, with a knowing expression of, "If you asked me nicely, then I would do as you ask".

As a general rule, though, when away from their "safe" environment my dogs have been pretty unquestioning in their obedience. Robs and I took this even further. I was once at a Christening where I observed with great interest, and not a little envy, that some clever owner had been able to leave the boot of their car open, with their Golden Retriever simply immovable through utter astounding obedience. I considered this to be a very useful habit to cultivate and made a mental note to train Robbie to stay in the boot of the car when in a foreign environment.

He learnt to do this brilliantly, and it was indeed a convenient behaviour sometimes. I was quite proud of this skill until I learnt the truth about the dog at the Christening.

A few years later I met the owners of that paragon of virtue again. I commented on the their dog's obedience and said I had admired it so much. I also took the opportunity to thank them for giving me the idea. They roared with laughter: the dog only stayed in the boot because she was too scared to get out! Oh, well, the boot trick is something Ellie is also expected to adhere on occasion, so I guess envy can also be the mother of invention too!

Other more cussed traits include: tantrums, sulks and sheer bloody-mindedness, the avowed intention to make sure that you know that, "I am not a dog, I'm a lurcher". This combined with an air of massive self-importance despite the will to please and, on occasion, it is never quite clear, which trait is to take the upper hand.

This is often expressed in their insistence in using the FRONT Door, and not the tradesman's entrance at any house, however grand. Actually the more stately the home, the greater their determination. This is not confined to my lurchers alone, however. I once read in Country Life of a lurcher called Frances, a rescue dog, probably of similar humble origins, though I am not sure, who also was determined that the only entrance to make was through the front door. Again, however stately the home, actually, the more stately the home, the more the determination to use the FRONT door.

As we know, this air of entitlement is not confined to houses, but to cars also, and other places – they possess an innate sense of superiority, of rightness and belonging that I have not witnessed in many other breeds, but their charming and outgoing manner allows them to get away with it.

Robbie had, and Ellie generally has, exquisite manners. When asked to say "*thank you*" to someone, they always comply. They, without fail seek you out, to thank you for their dinner but Robbie would go one further.

On leaving some friends, who had very reluctantly had him to stay with me when I was visiting, I asked Robs to go and say thank you to our host, calling him by name. There were quite a few people in the room, but Robs went unfailingly straight up to my friend and nodded his head and wagged his tail. He smiled in that peculiar lurcher manner, and marched back to me, as if to say, "Done it, can we go home now?", leaving behind a stunned silence in the room.

Robbie had a lovely smile and a great sense of humour; Paul Hogarth, the artist, once said of him, "I should paint him as The Laughing Cavalier". I wish he had for it was one of the qualities which I so cherished in my boy, ever the practical joker he would do anything for a laugh. Ellie loves people laughing but Robs had a rare comedic talent to actually entertain.

CHAPTER 52

So What Is A Lurcher?

A Loveable, Contrary, Clever, Cunning, Speedy, Loyal, Decisive, Beautiful Couch Potato a.k.a. Bed-Hogger.

Yes, these are the defining contradictions that characterise the lurcher group of dogs remains unrecognised as a breed by the Kennel Club, as they defy definition, and so the term"Lurcher" is merely descriptive. Wherein, perhaps, lies their not inconsiderable charm. I commonly hear it said, and I say it myself, that once you have lived with a lurcher then you will never go back and my husband, Christopher, is a testament to this.

Frequently asked the question, "What IS a lurcher?" I usually reply that it is generally accepted that they are a combination of a working and a hunting dog. But even that definition does not work, because Ellie's parents were both lurchers, though her grandparents were Deerhound, Greyhound, Collie and Bedlington terrier.

The best classification of lurchers I could find goes as follows:

"A cross-bred dog which is the offspring of a sighthound, mated with another breed, commonly a pastoral dog, such as a collie, retriever, sheepdog, or a terrier."

The archaic definition is: "a prowler, a swindler or a petty thief" derived from the Romany words "lur" for thief, and "cur" for mixed breed. Or possibly from the Middle English "lorchen" which means "to lurk."

(I personally disagree with this, because none of my lurchers have ever been thieves, and I would argue that these characteristics have been transposed by characters who wish to deflect the spotlight being shone on their own nefarious characters, and not those inherent in their dogs.)

The noted breeder, Brian Plummer, regards the Norfolk Lurcher to be the predecessor of the modern lurcher. Robbie and Ellie were both Bedlington crosses, also known as the Norfolk, which I think exhibit the best qualities of all the lurcher varieties.

What obfuscates the definitions is that, because lurchers are bred from such a genetically varied pool, you are never quite sure what is going to come out in the mix when you breed them. This leads to an interesting conundrum for the prospective owner: do you get a rescue dog of which there are many, too many, though not often the top notch type? Or do you get a puppy and hope for the best? I particularly adore the Bedlington crosses for the reasons that I have outlined extensively elsewhere, so I would want to be VERY sure about the outcome of any prospective marriage, as it were.

Anyway to the points of agreement, the Norfolk Lurcher generally exhibits:

- Great intelligence and speed

- Cunning
- Resourcefulness
- Greyhound shape with short or broken (long) coat, collie eyes, alert ears and slim defined waists.
- Height: 22"-26"
- Weight: 20kg-32kg
- Life expectancy of around 13 years.
- Obedience (within reason, theirs of course)

Myth has it that between 1400 and 1650, the English and Scottish governments banned commoners from owning sighthounds (wolfhounds, deerhounds or greyhounds), so lurchers may have been bred to circumvent any such legislation. However, there is no evidence of any such laws and I think it much more likely that the first lurchers were a happy accident of a local love affair that turned out happily ever after. When some observant person, probably about 8,000 years ago in Abyssinia, recognised the singular qualities of cross breeds and encouraged further breeding. Moreover if lurchers or a similar breed appeared in the Book of Kells, which was produced between the 6th and the 9th century, then lurchers must have appeared much earlier than the 14th century in the British Isles.

We know that the "dogification" (my word "meaning" civilisation and training by dogs) of humans began some 30,000 years ago, probably in South East Asia, when grey wolves are believed to have learnt to hunt alongside humans. If this is the case, then the self-reliance and decisiveness that some of these hounds exhibit, is actually an atavistic trait which has served them well over thousands of years. I have always firmly believed that the dog, in some way, chooses us, as memorably demonstrated by Ellie and Robbie. Two dogs who made it abundantly clear, at our very first meeting, that they were part of the decision process.

Backing up this supposition has been the recent discovery of the now almost infamous "Tooth" dug up at Blick Mead in Wiltshire which it is claimed belonged to a dog who travelled from Yorkshire with a group of people some 8,000 years ago. I say infamous, for as exciting as this discovery is, I fear there may be a lot of conclusions drawn hastily here.

However, disbelief suspended, researchers at Durham University have used carbon dating to discover the age of the tooth, and isotope analysis on the enamel revealed that the dog, which is tentatively described as of "Alsatian type", was born in the York area. The dog was drinking water from that area when young. He was part of a group who journeyed some 250 miles to Stonehenge, eating the same food as them. Scientific analysis demonstrated that other bones found near the tooth suggest a diet of salmon, trout, pike, wild pig and red deer. This is, of course, entirely analogous with not giving your dog tinned convenience food unless otherwise impossible.

The point that is exciting here, is that a wolf would not stray 250 miles from its territory, it is clearly the behaviour of a tamed, if not domesticated hound, who was in some sense part of the family, benefiting from the relationship. So nothing has changed over millennia!

CHAPTER 53

SPEED

And here we come to the subject of speed and a surprising fact: it is well known that the Cheetah is the fastest land animal on the planet, and the Falcon, the fastest bird with speeds of over 240 mph in a dive. But, it all depends on the measurements and the distance being measured. Cheetahs and horses are blessed with fantastic turns of speed at a pace of up to 70 mph and 45 mph at a sprint respectively, though the cheetah can only sustain this speed over 200-300 yards, and the horse up to a quarter-mile.

The greyhound, however, can reach 45 mph in 6 strides from a standing start; moreover he is also in it for the long haul, being able to sustain swiftness of over 35 mph for over seven miles.

Birds, however, can reach much higher speeds: we celebrate "swiftness" when the bird of that name can reach 106 mph in direct flight, the Peregrine Falcon 69 mph in direct pursuit, but up to an incredible 240mph in a dive to prey.

This propensity for speed has been bred specifically into sighthounds, such as the greyhounds, whippets, salukis and the Afghans, whose job is to spot quarry and run it down.

Their huge chest and oversized heart allow them to circulate large amounts of oxygen when at the sprint, and their narrow waist allows them to bend their body, so that every stride carries them further than their body length and, even more importantly, to turn at extreme speeds.

Is it any accident that the fastest animals are often the most admired? More to the point, a dog that is capable of reaching such extreme speeds also needs to be biddable, otherwise they would never ever return from such epic distances.

Here is another potential reason for the cross-breeding: sighthounds may be very speedy but they are not known for their intelligence nor their obedience (apologies to any owners, but Ferdie was a Deerhound cross and obedience and intelligence were far from being her greatest qualities, though her affectionate nature and her massive persistence in the pursuit of deer were outstanding).

A canny person might easily have contrived the union of the pastoral working dog with the sighthound in order to harness the best qualities of each, sacrificing just a little speed to acquire an animal that might not only return home with the quarry, but might, just might, be willing to listen to his human companion.

CHAPTER 54

Other Considerations

Personally, I do not think lurchers make a great town dog, though some of the breeds from which they descend do, such as the Bedlington, or Greyhound, due to their sofa hugging qualities. Though, there are number of my acquaintances who have had a very successful urban relationship with a Lurcher, but only when they have access to some large and dog friendly green area for exercise such as Richmond Park.

They frequently show similar physical traits to their prey, the turning speed of the hare, the long legged doe eyes of the deer; Ferdie, the deer-hound cross had a rangy loping trot which reminded me of nothing less than a big cat, she always looked as if she belonged on safari.

They require at least one supervised walk a day, somewhere to show off their extreme speed. Expect vet's bills (massive), more than a few stitches, insurance is a necessity, as is preferably a cat free zone, as though they can forge good relationships with family cats, they always seem to deem the neighbours' cats fair game.

All of ours have shown great affinity for the rugged outdoors, but also for the finer things in life such as hair brushing, beds, parties and all the other little luxuries of life. Also, they have the most peculiar habit of turning up their noses at food that does not appeal nor please them, particularly when they are not hungry. Ellie has even been known to go hungry, if she is displeased with our choice of dinner.

In short, if the above appeals to you, and you can weather their infrequent but dramatic accidents, they are the ideal dog. (And I am not biased at all…!)

EPIDOGUE

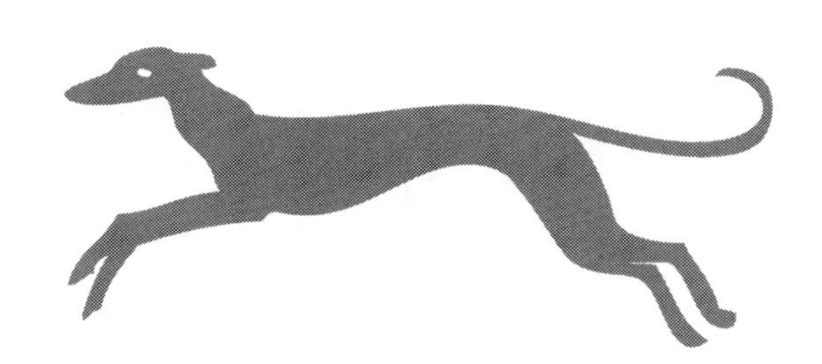

River Dogs - Sea Dogs

We have now another river dog: Ellie loves the very same shore along the River Severn. We took her down there a few years ago, and spent a morning exploring the beach I played on as a child. She is also a sea dog - she adores her beach on the Gower Peninsular, and eagerly anticipates our occasional visits there.

As I write this in Autumn, the canine mudlark lies beside me on the bed. At 6:25am comes the first herald of the dawn, the little blackbird squawk, and I draw back the curtain to see the grey light of morning breaking over the trees on the hill. My little muse is snoring, may be even dreaming on her times upon the shores.

We have wonderful photos of that first day at the River Severn, where in the salt marsh she explored the unfamiliar smells of the river side, trotted through the really ancient rocks and fossils with more interest in their texture, than their geological importance, leapt over boulders and grasses, and raced in one of the channels between the rocks almost under the bridge which is safe at low tide.

There is a tiny outcrop that serves as a small island, where you can go at very low tide, and we scrambled over the slippery seaweed clad rocks and splashed through salty shallows. It was, we think, the first time she had encountered the startling unfamiliarity of the salt spray as it splattered her coat. She made the entirely natural and uninformed mistake of trying to drink it. Shaking her head in disgust, she stares at us as if to say, "Why didn't you tell me?" before bouncing off again in the trickling and insane apparent safety of low tide.

Now she adores splashing around in the sea on the Welsh coast and spends hours playing on and investigating her beach. These are the enduring memories of my most beloved dog.

On Dogs Generally

(And Lurchers Particularly)

Dogs are not only our companions on walks, by the fire, in sickness and in health, but also grace our walls and our literature. Truly, the dog is man's most constant and faithful and loving companion – though who is to say who chose who? And on what basis?

I, for one was first enchanted by their looks, their grace, their speed, and then, began to appreciate the astonishing complexity of these animals. They can run faster, jump higher, decipher complex escape problems, show determination beyond sometimes the bounds of possibility, kindness and compassion without reward, loyalty and love without question, more than any human I have yet encountered.

Ellie has brought together the superlative qualities of all the dogs whom I have had the privilege to know well. I balk at the word own, for I truly do not know either who owns or indeed chose whom.

Ellie is certainly a case in point, she clearly rewrote the rule book.

I watch her now just past mid-day basking in the October sun, warm inside, cool outside, and I continue to ponder that ineffable question.

Surely, I have mused, dogs know they are on to a good thing – food, walkies, warmth, bed and loving care; they give us in return something absolutely priceless,

They teach us to see, lead us outside, coax us to care, encourage us to listen, to wonder at their different perspective and view of the world, but above all

THEY OPEN OUR HEARTS, our ears and our minds.

To say nothing about that all important sense of smell...

On the Death of a Dog

Goodbye dearest friend, my love, my heart's home,
Never no more shall I see you roam,
Except in the mind's eye,
Around the home,
Where now I sit,
And all alone.
I so wish I could give you one final bone.

The love I have felt for the animals with whom I have been lucky enough to share my life with, so far, has been so deep, so wide, so tender, I hesitate to label what is ineffable and universal.

They have been my "tapestry hounds" drawing together the threads of my existence and weaving my life into something so much greater than its component parts. Dogs, generally, and Lurchers, in particular, have and continue to elevate daily life from the mundane to the sublime.

So untainted by worldly desires, (other than the next treat), so inspired by the world they experience around them: grass, air, sky, the ground beneath their feet or paws or hoofs, thunders in celebration of the joy of just being.

They know the meaning of enough: a good walk, a jolly good supper, preferably with pre-prandial nap, perhaps, nowadays a nice bit of telly, a comfortable bed, and a sleep, when the long trick's over.

One day, that sleep will be uninterrupted, but the dream will go on. The day we first took Ellie to the beach by the River Severn, was a day when it seemed all my happiest childhood memories, and those of Benjie and Robbie, and Amie and Fly and Ferdie came together. The shadow dogs all played together with Ellie, all sea dogs, river dogs, many dogs, running along the shore. And all my friends go coursing, coursing with the stars through the night sky, as I dream of yesterday and tomorrow: my spirit dogs, my beautiful Ellie, and future friends as yet unmet.

But, for now, there are still adventures to be retold, relived and still to come.

THE END OR IS IT JUST THE BEGINNING?

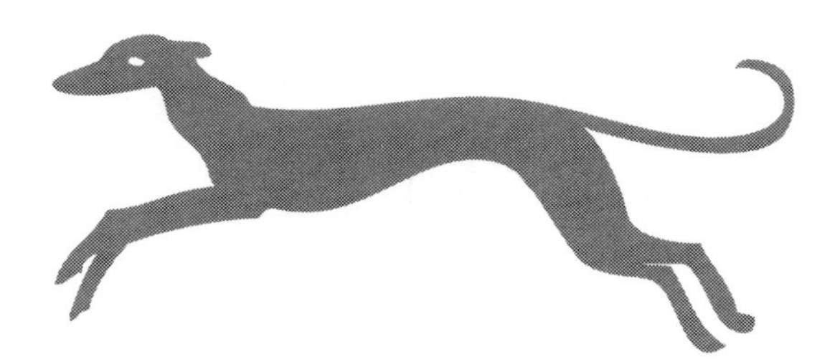

ACKNOWLEDGEMENTS

I cannot thank those that have helped me enough for their quiet and loving support, including Darling Jilly for her kind suggestions; dearest Pamela for her skilled and on-going help and Uncle George for putting up with having to listen so much. Pam, who recently published her own book, Morag for her foresight, and my parents for allowing us to grow up with so many animals.

Not forgetting my wonderful father-in-law for his unswerving encouragement and support, to say nothing of my husband's patience with the numerous revisions and his skill in putting together the finished work.

Last, but not least, a massive thank you to all those animals with whom I have been privileged to share my life and most especially to my darling lurchers, Robbie and Ellie, without whom this book would not exist.

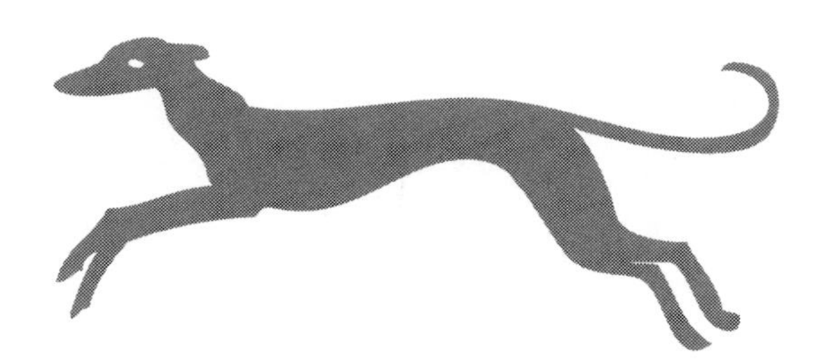

More Information

You can find more information about and pictures of the dogs and other animal characters in this book on our dedicated website. Links to her YouTube videos can be found on www.booksaboutdogs.co.uk.

WWW.BOOKSABOUTDOGS.CO.UK

You can find full colour versions of the pictures in this book, together with others, on this website

WWW.ELLIEWHITE.CO.UK

WWW.INSTAGRAM.COM/THELURCHERLASS

About The Author

Pippa started her working life as an analyst for JP Morgan, having read English Literature at Cambridge University.

She has since established a number of luxury home accessories businesses, both retail and trade.

Pippa has published two children's books on Amazon, through her publishing company, Ellie White. Ellie White is her first non-fiction book and the first in a series of books about dogs and other animals.

Trained as an NLP therapist, she is also actively involved in Charity Fundraising, and has volunteered in teaching drama at local schools.

Pippa has one grown up daughter and lives in Gloucestershire with her husband, and of course, Ellie. Her interests are widespread and include writing, music, interior design, cookery, organic lifestyle and the countryside, and naturally, lurchers, horses and other animals.

Printed in Poland
by Amazon Fulfillment
Poland Sp. z o.o., Wrocław